ENVIRONMENTAL SCIENCE, ENGINEERING AND TECHNOLOGY

USE OF BIOLOGICAL ASSESSMENTS TO SUPPORT WATER QUALITY MANAGEMENT

A PRIMER

Environmental Science, Engineering and Technology

Additional books in this series can be found on Nova's website under the Series tab.

Additional E-books in this series can be found on Nova's website under the E-book tab.

Water Resource Planning, Development and Management

Additional books in this series can be found on Nova's website under the Series tab.

Additional E-books in this series can be found on Nova's website under the E-book tab.

ENVIRONMENTAL SCIENCE, ENGINEERING AND TECHNOLOGY

USE OF BIOLOGICAL ASSESSMENTS TO SUPPORT WATER QUALITY MANAGEMENT

A PRIMER

IAN W. SORENSON
EDITOR

nova publishers
New York

For permission to use material from this book please contact us:
Telephone 631-231-7269; Fax 631-231-8175
Web Site: http://www.novapublishers.com

Library of Congress Cataloging-in-Publication Data

ISBN: 978-1-62948-481-5

Published by Nova Science Publishers, Inc. † New York

CONTENTS

Preface		**vii**
Chapter 1	A Primer on Using Biological Assessments to Support Water Quality Management *U.S. Environmen1tal Protection Agency*	**1**
Chapter 2	Clean Water Act: A Summary of the Law *Claudia Copeland*	**123**
Index		**139**

PREFACE

In this book, the role of biological assessments in a variety of water quality management program applications is discussed, including reporting on the condition of aquatic biota, establishing biological criteria, and assessing the effectiveness of Total Maximum Daily Load determinations and pollutant source controls. This book provides a brief discussion of technical tools and approaches for developing strong biological assessment programs and presents examples of successful application of those tools. The objective of the Clean Water Act (CWA), and water quality management programs generally, is "to restore and maintain the chemical, physical, and biological integrity of the Nation's waters." Although we have achieved major water quality improvements over the past four decades and have reduced the discharge of many toxic chemicals into our nation's waters, many environmental challenges remain, such as loss and fragmentation of habitat, altered hydrology, invasive species, climate change, discharge of new chemicals, stormwater, and nitrogen or phosphorus (nutrient) pollution.

Measuring the condition of the resident biota in surface waters using biological assessments and incorporating that information into management decisions can be an important tool to help federal, state, and tribal water quality management programs meet many of their challenges.

Chapter 1 – This guide serves as a primer on the role of biological assessments in a variety of water quality management program applications, including reporting on the condition of the aquatic biota, establishing biological criteria, and assessing the effectiveness of Total Maximum Daily Load determinations and pollutant source controls. This guide provides a brief discussion of technical tools and approaches for developing strong biological assessment programs and presents examples of successful application of those tools.

The objective of the Clean Water Act (CWA), and water quality management programs generally, is "to restore and maintain the chemical, physical, and biological integrity of the Nation's waters." Although we have achieved major water quality improvements over the past four decades and have reduced the discharge of many toxic chemicals into our nation's waters, many environmental challenges remain, such as loss and fragmentation of habitat, altered hydrology, invasive species, climate change, discharge of new chemicals, stormwater, and nitrogen or phosphorus (nutrient) pollution. In the face of such challenges, how can we best deploy our water quality programs to meet the vision of the CWA for protection of aquatic life?

Measuring the condition of the resident biota in surface waters using biological assessments and incorporating that information into management decisions can be an important tool to help federal, state, and tribal water quality management programs meet many of the challenges. Biological assessments are an evaluation of the condition of a waterbody using surveys of the structure and function of a community of resident biota (e.g., fish, benthic macroinvertebrates, periphyton, amphibians) (for more information, see *Biological Assessment Key Concepts and Terms*)[1]. Assessments of habitat condition, both instream and riparian, are typically conducted simultaneously. Such information can reflect the overall ecological integrity of a waterbody and provides a direct measure of both present and past effects of stressors on the biological integrity of an aquatic ecosystem. The benefit of a biological assessment program is based in its capability to:

- Characterize the biological condition of a waterbody relative to water quality standards (WQS).
- Integrate the cumulative effects of different stressors from multiple sources, thus providing a holistic measure of their aggregate effect.
- Detect aquatic life impairment from unmeasured stressors and unknown sources of impairment.
- Provide field data on biotic response variables to support development of empirical stressor response models.
- Inform water quality and natural resource managers, stakeholders, and the public on the environmental outcomes of actions taken.

It is EPA's long-standing policy that biological assessments should be fully integrated in state and tribal water quality programs and used together with whole effluent and ambient toxicity testing, and with chemical-specific analyses, to assess attainment of designated aquatic life uses in WQS (USEPA 1991b). Each of these methods can be used to provide a valid assessment of aquatic life use impairment. Biological assessments complement chemical-specific, physical, and whole effluent toxicity measures of stress and exposure by directly assessing the response of the community in the field (USEPA 1991a). Measurable changes in the biotic community—for example, the return of native species, decrease in anomalies and lesions in fish and amphibians, and decrease in pollution-tolerant species paired with an increase in pollution-sensitive species—can be readily communicated to the public and the regulated community. This can result in greater stakeholder understanding of effects from stressors and support for management actions. Additionally, as response-stressor relationships are documented, biological assessments in concert with stressor data can be used to help predict and track environmental outcomes of management actions.

Chapter 2 – The principal law governing pollution of the nation's surface waters is the Federal Water Pollution Control Act, or Clean Water Act. Originally enacted in 1948, it was totally revised by amendments in 1972 that gave the act its current dimensions. The 1972 legislation spelled out ambitious programs for water quality improvement that have since been expanded and are still being implemented by industries and municipalities.

This report presents a summary of the law, describing the statute without discussing its implementation.

The Clean Water Act consists of two major parts, one being the provisions which authorize federal financial assistance for municipal sewage treatment plant construction. The other is the regulatory requirements that apply to industrial and municipal dischargers. The act has been termed a technology-forcing statute because of the rigorous demands placed on those who are regulated by it to achieve higher and higher levels of pollution abatement under deadlines specified in the law. Early on, emphasis was on controlling discharges of conventional pollutants (e.g., suspended solids or bacteria that are biodegradable and occur naturally in the aquatic environment), while control of toxic pollutant discharges has been a key focus of water quality programs more recently.

Prior to 1987, programs were primarily directed at point source pollution, that is, wastes discharged from discrete sources such as pipes and outfalls. Amendments to the law in that year authorized measures to address nonpoint source pollution (runoff from farm lands, forests, construction sites, and urban areas), which is estimated to represent more than 50% of the nation's remaining water pollution problems. The act also prohibits discharge of oil and hazardous substances into U.S. waters.

Under this act, federal jurisdiction is broad, particularly regarding establishment of national standards or effluent limitations. Certain responsibilities are delegated to the states, and the act embodies a philosophy of federal-state partnership in which the federal government sets the agenda and standards for pollution abatement, while states carry out day-to-day activities of implementation and enforcement.

To achieve its objectives, the act is based on the concept that all discharges into the nation's waters are unlawful, unless specifically authorized by a permit, which is the act's principal enforcement tool. The law has civil, criminal, and administrative enforcement provisions and also permits citizen suit enforcement.

Financial assistance for constructing municipal sewage treatment plants and certain other types of water quality improvements projects is authorized under title VI. It authorizes grants to capitalize State Water Pollution Control Revolving Funds, or loan programs. States contribute matching funds, and under the revolving loan fund concept, monies used for wastewater treatment construction are repaid to states, to be available for future construction in other communities

In: Use of Biological Assessments ...
Editor: Ian W. Sorenson
ISBN: 978-1-62948-481-5

Chapter 1

A PRIMER ON USING BIOLOGICAL ASSESSMENTS TO SUPPORT WATER QUALITY MANAGEMENT[*]

U.S. Environmental Protection Agency

DISCLAIMER

The discussion in this document is intended solely to provide information on advancements in the field of biological assessments and on use of biological assessments to support state water quality management programs. The statutory provisions and the U.S. Environmental Protection Agency (EPA) regulations described in this document contain legally binding requirements. This document is not a regulation itself, nor does it change or substitute for those provisions or regulations. The document does not substitute for the Clean Water Act, a National Pollutant Discharge Elimination System permit, or EPA or state regulations applicable to permits; nor is this document a permit or regulation itself. Thus, it does not impose legally binding requirements on EPA, states, tribes, or the regulatory community. This document does not confer legal rights or impose legal obligations on any member of the public.

---

[*] This is an edited, reformatted and augmented version of U.S. Environmental Protection Agency, Publication No. EPA 810-R-11-01, dated October 2011.

While EPA has made every effort to ensure the accuracy of the discussion in this document, the obligations of the regulated community are determined by statutes, regulations, and other legally binding requirements. In the event of a conflict between the discussion in this document and any statute or regulation, this document will not be controlling.

The general descriptions provided here might not apply to a situation depending on the circumstances. Interested parties are free to raise questions and objections about the substance of this document and the appropriateness of the application of the information presented to a situation. EPA and other decision makers retain the discretion to adopt approaches on a case-by-case basis that differ from those described in this document where appropriate.

Mention of any trade names, products, or services is not and should not be interpreted as conveying official EPA approval, endorsement, or recommendation.

This is a living document and might be revised periodically. EPA could revise this document without public notice to reflect changes in EPA policy, guidance, and advancements in field of biological assessments. EPA welcomes public input on this document at any time. Send comments to Susan Jackson, Office of Science and Technology, Office of Water, U.S. Environmental Protection Agency, 1200 Pennsylvania Avenue, Mail Code 4304T, Washington, DC 20460.

The reference section and Appendix A of this document were updated February 2012 to include publication dates for USEPA technical documents and website that were originally cited as draft.

FOREWORD

This guide serves as a primer on the role of biological assessments in a variety of water quality management program applications, including reporting on the condition of the aquatic biota, establishing biological criteria, and assessing the effectiveness of Total Maximum Daily Load determinations and pollutant source controls. This guide provides a brief discussion of technical tools and approaches for developing strong biological assessment programs and presents examples of successful application of those tools.

The objective of the Clean Water Act (CWA), and water quality management programs generally, is "to restore and maintain the chemical, physical, and biological integrity of the Nation's waters." Although we have achieved major water quality improvements over the past four decades and have reduced the discharge of many toxic chemicals into our nation's waters, many environmental challenges remain, such as loss and fragmentation of habitat, altered hydrology, invasive species, climate change, discharge of new chemicals, stormwater, and nitrogen or phosphorus (nutrient) pollution. In the face of such challenges, how can we best deploy our water quality programs to meet the vision of the CWA for protection of aquatic life?

Biological integrity has been defined to mean the capability of supporting and maintaining a balanced, integrated, and adaptive community of organisms having a composition and diversity comparable to that of natural habitats of the region (Frey 1975; modified by Karr and Dudley 1981). **Biological assessments** can be used to directly measure the condition of the biota residing in a waterbody and provide information on biological integrity. Resident biota include species that spend all or a part of their life cycle in the aquatic environment.

Measuring the condition of the resident biota in surface waters using biological assessments and incorporating that information into management decisions can be an important tool to help federal, state, and tribal water quality management programs meet many of the challenges. Biological assessments are an evaluation of the condition of a waterbody using surveys of the structure and function of a community of resident biota (e.g., fish, benthic macroinvertebrates, periphyton, amphibians) (for more information, see *Biological Assessment Key Concepts and Terms*)[1]. Assessments of habitat condition, both instream and riparian, are typically conducted simultaneously. Such information can reflect the overall ecological integrity of a waterbody and provides a direct measure of both present and past effects of stressors on the biological integrity of an aquatic ecosystem. The benefit of a biological assessment program is based in its capability to:

- Characterize the biological condition of a waterbody relative to water quality standards (WQS).
- Integrate the cumulative effects of different stressors from multiple sources, thus providing a holistic measure of their aggregate effect.
- Detect aquatic life impairment from unmeasured stressors and unknown sources of impairment.
- Provide field data on biotic response variables to support development of empirical stressor response models.

- Inform water quality and natural resource managers, stakeholders, and the public on the environmental outcomes of actions taken.

It is EPA's long-standing policy that biological assessments should be fully integrated in state and tribal water quality programs and used together with whole effluent and ambient toxicity testing, and with chemical-specific analyses, to assess attainment of designated aquatic life uses in WQS (USEPA 1991b). Each of these methods can be used to provide a valid assessment of aquatic life use impairment. Biological assessments complement chemical-specific, physical, and whole effluent toxicity measures of stress and exposure by directly assessing the response of the community in the field (USEPA 1991a). Measurable changes in the biotic community—for example, the return of native species, decrease in anomalies and lesions in fish and amphibians, and decrease in pollution-tolerant species paired with an increase in pollution-sensitive species—can be readily communicated to the public and the regulated community. This can result in greater stakeholder understanding of effects from stressors and support for management actions. Additionally, as response-stressor relationships are documented, biological assessments in concert with stressor data can be used to help predict and track environmental outcomes of management actions.

SECTION 1. INCORPORATING BIOLOGICAL ASSESSMENTS INTO WATER QUALITY MANAGEMENT

1.1. Why Is Measuring Biological Condition Important?

With the passage of the Clean Water Act (CWA) in 1972 and subsequent national investment in water infrastructure and regulation, much work has been done to restore rivers, lakes, streams, wetlands, and estuaries. However, despite our best efforts and many documented successes, we continue to lose aquatic resources (Figure 1-1) (H. John Heinz III Center for Science, Economics, and the Environment 2008; Jelks et al. 2008; USEPA 2006). Pollutants (e.g., pathogens, metals, nitrogen, phosphorus pollution) continue to be major causes of water quality degradation. Additionally, the impact of other significant stressors, including habitat loss and fragmentation, hydrologic alteration, invasive species, and climate change, can be better understood using analytical tools and information that can operate at the ecosystem scale, such as biological assessments.

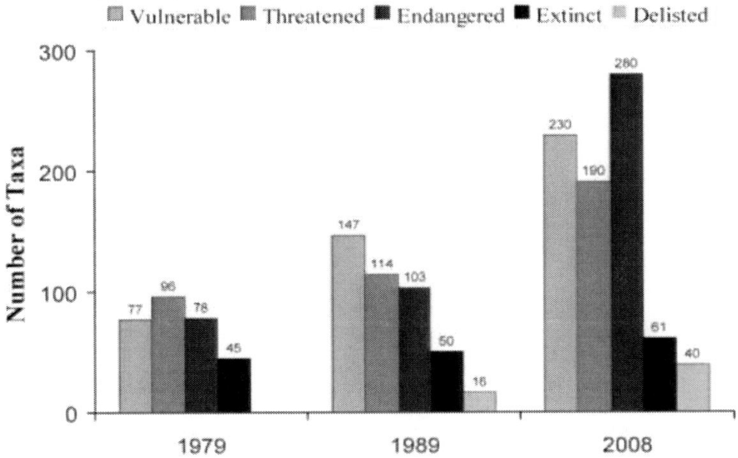

Source: Jelks et al. 2008

Note: The increase in total number of taxa identified as vulnerable, threatened, or endangered might be due in part to improvements in our understanding, naming, and assessing aquatic resources, resulting in more complete and accurate assessments.

Figure 1-1. Numbers of imperiled North American freshwater and diadromous fish taxa.

Biological assessments can be used to directly measure the overall biological integrity of an aquatic community and the synergistic effects of stressors on the aquatic biota residing in a waterbody where there are well-developed biological assessment programs (Figure 1-2) (USEPA 2003). Resident biota function as continual monitors of environmental quality, increasing the sensitivity of our assessments by providing a continuous measure of exposure to stressors and access to responses from species that cannot be reared in the laboratory. This increases the likelihood of detecting the effects of episodic events (e.g., spills, dumping, treatment plant malfunctions), toxic nonpoint source (NPS) pollution (e.g., agricultural pesticides), cumulative pollution (i.e., multiple impacts over time or continuous low-level stress), nontoxic mechanisms of impact (e.g., trophic structure changes due to nutrient enrichment), or other impacts that periodic chemical sampling might not detect. Biotic response to impacts on the physical habitat such as sedimentation from stormwater runoff and physical habitat alterations from dredging, filling, and channelization can also be detected using biological assessments.

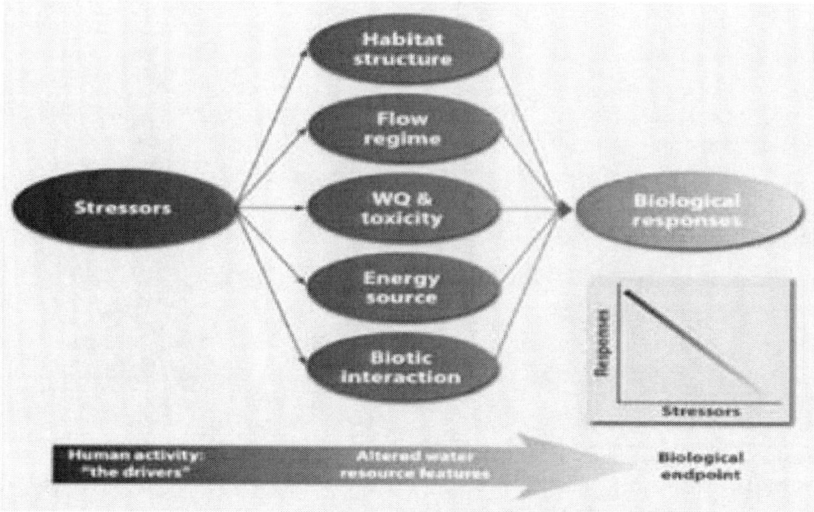

Figure 1-2. Biological assessments provide information on the cumulative effects on aquatic communities from multiple stressors. Figure courtesy of David Allen, University of Michigan.

States and tribes have used biological assessments to set environmental goals, detect degradation, prioritize management actions, and track improvements (USEPA 2002). Multiple examples of applications are presented in Section 3. Additionally, the U.S. Environmental Protection Agency (EPA) [2] and U.S. Geological Survey (USGS) [3] are conducting national and regional assessments of the condition of aquatic communities and the presence and distribution of stressors that affect the aquatic biota. The EPA National Aquatic Resource Surveys (NARS) program employs a probability-based sampling design while the USGS National Water-Quality Assessment (NAWQA) Program utilizes a targeted design. The data provide a baseline for assessing biological conditions and key stressors over time and tracking environmental improvements at the national or regional level (Figure 1-3).

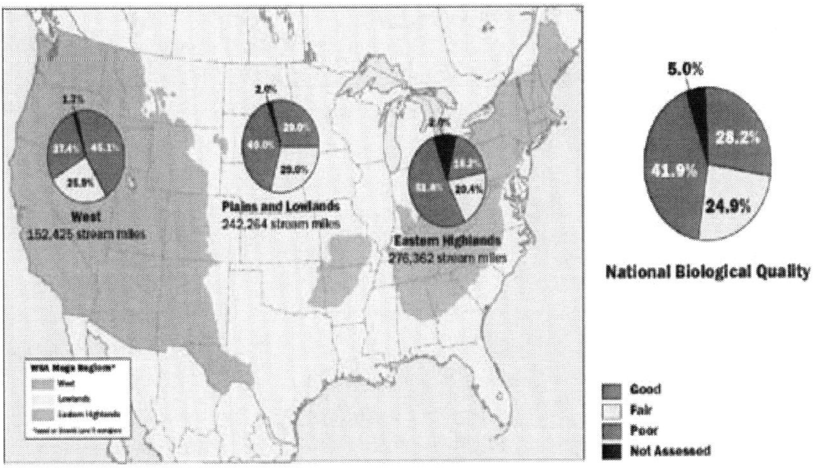

Source: USEPA 2006.

Figure 1-3. Biological condition of our nation's streams. In its first survey of stream condition, EPA found that 28 percent of the nation's stream miles are in good condition compared to the best existing reference sites in their regions, 25 percent are in fair condition, and 42 percent are in poor condition.

1.2. Using Biological Assessment Information in State and Tribal Water Quality Management Programs

Biological assessment information has been used by states and tribes to:

- ***Define goals for a waterbody***—Information on the composition of a naturally occurring aquatic community can provide a description of the expected biological condition for other similar waterbodies and a benchmark against which to measure the biological integrity of surface waters. Many states and tribes have used such information to more precisely define their designated aquatic life uses, develop *biological criteria*, and measure the effectiveness of controls and management actions to achieve those uses.
- ***Report status and trends***—Depending on level of effort and detail, biological assessments can provide information on the status of the condition of the expected aquatic biota in a waterbody and, over time with continued monitoring, provide information on long-term trends.

- ***Identify high-quality waters and watersheds***—Biological assessments can be used to identify high-quality waters and watersheds and support implementation of state and tribal antidegradation policies.
- ***Document biological response to stressors***—Biological assessments can provide information to help develop biological response signatures (e.g., a measurable, repeatable response of specific species to a stressor or category of stressors). Examples include sensitivity of mayfly species (pollution-sensitive aquatic insects) to metal toxicity or temperature-specific preferences of fish species. Such information can provide an additional line of evidence to support stressor identification and causal analysis (USEPA 2000a), as well as to inform numeric criteria development (USEPA 2010a).
- ***Complement pollutant-specific ambient water quality criteria***—Biological assessment information can complement water quality standards (WQS) by providing field information on the cumulative effects on aquatic life from multiple pollutants, as well as detecting impacts from pollutants that do not have EPA recommended numeric criteria.
- ***Complement direct measures of whole effluent toxicity (WET) tests***—Biological assessments can provide information to help document improvements in aquatic life following actions taken to address the aggregate toxic effects of wastewater discharge effluents detected through laboratory WET tests. Additionally, biological assessments complement WET tests by directly measuring the cumulative or post-impact effects that both point source and NPS contaminants have on aquatic biota in the field.
- ***Address water quality impacts of climate change***—EPA, states, and tribes are exploring how biological assessments can be used in concert with physical, chemical, and land use data to help identify baseline biological conditions against which the effects of global climate change on aquatic life can be studied and compared. Such information could enable a water quality management program to calibrate biological assessment endpoints and criteria to adjust for long-term climate change conditions. Additionally, long-term data sets will enable trends analysis and support predictive modeling and forecast analysis.

1.3. Water Quality Program Applications and Case Studies

The CWA employs a variety of regulatory and nonregulatory approaches to reduce direct pollutant discharges into waterways, finance municipal wastewater treatment facilities, and manage polluted runoff. Those approaches are employed to achieve the broader goal of restoring and maintaining the chemical, physical, and biological integrity of the nation's waters. The role of biological assessment information to support such approaches is described below, and case studies of successful implementation are provided in Section 3.

Water Quality Standards

State and tribal WQS programs can use biological assessment information in developing descriptions of CWA-designated aquatic life uses in terms of the expected biological community. For example, in states and tribes that identify high-quality waters for antidegradation purposes on a waterbody-by-waterbody basis, biological assessments can provide information to help define and protect existing aquatic life uses and identify Tier 2 waters (e.g., where the quality of the waters exceed levels necessary to support propagation of fish, shellfish, and wildlife and recreation in and on the water) and Outstanding National Resource Waters (ONRWs). Maryland is using biological assessments to help identify high-quality streams for antidegradation purposes on a waterbody-by-waterbody basis (case study 3.3). Pennsylvania is exploring the use of biological assessment information to help assess attainment of aquatic life uses and to describe biological characteristics of waters along a gradient of condition (case study 3.4). This information may potentially be used to support protection of waters of the highest quality that require special protection. Arizona used biological assessments to develop numeric biological criteria using the reference condition approach (Stoddard et al. 2006) that takes into account the quality of the reference sites (case example 3.2).

Some states have calibrated biological response to gradients of anthropogenic stress impacting surface waters (see Section 2, Tool #2, *The Biological Condition Gradient*). This approach, when applied to WQS by defining the designated aquatic life uses along a gradient of condition, has provided these states with the capability to improve waters incrementally, protect high-quality waters, and help identify factors that affect attainability. For example, Maine assigns a waterbody to a specific condition class on the basis of its current condition and potential for improvement. Numeric biological criteria have been developed for each class and adopted into their WQS (case study 3.1). Over the past 30 years, the use designations for many streams and rivers in Maine have been upgraded according to documented biological improvements and attainment of the biological criteria that define higher quality use classes. This approach is sometimes referred to as tiered aquatic life uses and has also been implemented by the State of Ohio (case study 3.5).

Additionally, biological assessments can provide information on the species composition at a site under consideration for site-specific criteria. Using the species recalculation procedure, a state or tribe can adjust chemical water quality to reflect the chemical sensitivity of species that occur at a site (USEPA 1994). Biological assessment information may support modification of the default species sensitivity distribution to better reflect the expected community composition at the site. For example, if the site is a naturally occurring warm body of water, coldwater fish species could be replaced by resident warmwater fish species in the species sensitivity distribution from which a site-specific criterion is calculated.

Monitoring and Assessment

Biological monitoring and assessments provide data to aquatic resources managers at the local, state, tribal, regional, and national levels to help assess status and trends of aquatic resources as well as to measure the effectiveness of management actions to protect or restore waters. For example, the biological monitoring program in Montgomery County, Maryland, produces biological assessment information on the condition of the County's streams and the effectiveness of innovative best management practices (BMPs) for stormwater control.[4] At the state level, the Maryland Department of the Environment (MDE) conducts biological monitoring to evaluate permit effectiveness, conduct impact assessments, and identify high-quality waters (case studies 3.3 and 3.12). Also, Maryland Department of Natural Resources (MDNR)[5] provides MDE and the public with a statewide biological assessment of status and trends for streams and rivers that may serve as a yardstick for measuring the overall effectiveness of local and state management actions.

Biological assessment information has been used by counties and state/tribal agencies to facilitate collaboration and effective use of limited resources. For example, two state agencies in Oregon jointly conducted biological assessments to address their information needs (case study 3.17). For the Oregon Department of Fish and Wildlife (ODFW), monitoring of aquatic benthic macroinvertebrate communities in streams and rivers, in conjunction with chemical and physical monitoring, provided important information on water quality and habitat conditions identified as critical to coho salmon viability. Oregon's Department of Environmental Quality (ODEQ) used the same biological assessment information to assess attainment of the designated uses to protect and maintain salmonid populations.

At the national level, biological data from the *National Aquatic Resource Surveys*[6] are being used in EPA's strategic plan to track improvements in water quality for streams, rivers, wetlands, and coastal waters. The results of the first national surveys for streams and coastal waters are included in EPA's *Report on the Environment.*[7] These surveys, which incorporate a statistical probabilistic design, are key tools for communicating to the public what the Agency knows about the condition of the nation's waters at national and regional scales. The biological components of the national surveys will continue to provide nationally consistent indicators of water quality that can be used to gauge the overall effect of the national investment in protecting and restoring the nation's watersheds.

EPA also uses biological assessments to assess status and trends at a regional or large ecosystem scale (e.g., in the Upper Mississippi River Basin or the Great Lakes) and measure biological response to restoration efforts related to disasters (e.g., Hurricane Katrina and the Gulf of Mexico oil spill). National and regional biological assessments provide information that helps facilitate interagency collaboration for large-scale restoration and protection efforts. For example, a recent USGS multiregional assessment found that alteration of streamflow is a major predictor of biological integrity of both fish and macroinvertebrate communities (Carlisle et al. 2010). Alterations in stream flow are associated with riparian disturbance and can influence the release of nitrogen, phosphorus, and sediments into streams (Poff and Zimmerman 2010). The combined results of national, regional, and state/tribal ecological assessments will provide the data needed to predict and better manage future impacts of stressors from human activities such as urban development, water allocation, and agriculture. The results of different program actions to address different stressors and their sources can be related to a common measure of environmental improvement—the condition of the aquatic biota.

Identification of Impaired and Threatened Waters in States' Integrated Water Quality Reports

Under section 303(d) of the CWA and supporting regulations (40 CFR 130.7), states, territories, and authorized tribes (hereafter referred to as states) are required to develop lists of impaired and threatened waters that require Total Maximum Daily Loads (TMDLs). Impaired waters are those that do not meet any applicable WQS, including designated uses, narrative criteria and numeric criteria such as biological criteria adopted as a standard. EPA recommends that states consider as threatened those waters that are currently attaining WQS, but which are expected to not meet WQS by the next listing cycle (every 2 years). Consistent with EPA recommendation, many states consolidate their section 303(d) and section 305(b) reporting requirement into one "integrated" report.

If biological assessments indicate that a waterbody is impaired or threatened, the waterbody is included on the state's section 303(d) list and scheduled for TMDL development. Some 30 states have used biological assessment information as the basis for concluding that designated aquatic life use(s) were not supported and included these waters on their section 303(d) lists. In some cases, these listings were based on comparison of the biological assessments to state-adopted numeric biological water quality criteria. However, in most cases, biological assessments were treated as translations of one or more of a state's narrative water quality criteria or as direct evidence that designated aquatic life uses were not supported.

How to reconcile conflicting results among different datasets (e.g., chemical, physical, biological) is discussed in EPA's Integrated Reporting Guidance (IRG) for the 2006 sections 303(d) and 305(b) reporting cycle. Also discussed in the IRG, if a designated use, such as aquatic life, is not supported and the water is impaired or threatened, the fact that the specific pollutant may not be known does not provide a basis for excluding the water from the section 303(d) list.[8] These waters are often identified on a state's list as cause or pollutant unknown. These waters must be included on the list until the pollutant is identified and a TMDL completed or the state can demonstrate that no pollutant(s) cause or contribute to the impairment. For example, in 1998, Iowa listed a 20-mile segment of the North Fork Maquoketa River as aquatic life use impaired—cause unknown, based on biological assessments. Using EPA's CADDIS stressor identification (SI) methodology, Iowa determined that the aquatic life use was impaired due to sediments, nutrients, and ammonia (see Tool #3, Stressor Identification and Causal Analysis/Diagnosis

Decision Information System). A TMDL was developed for each of these pollutants and these were approved by EPA in 2007 (case study 3.7).

Development of Total Maximum Daily Loads

Under the CWA, states are required to develop TMDLs for impaired and threatened waters on their 303(d) lists. States and tribes may use biological assessments to support developing one or more water quality targets for the pollutant of concern on the basis of well-documented stressor-response relationships, from reference conditions or through use of mechanistic modeling. This is done in conjunction with other water quality monitoring data, such as data on concentrations of specific stressors and toxicity effects. For example, Connecticut has developed a relationship between pollutant loads, stormwater flows, and impervious land cover (IC) for streams in small watersheds with no other known point source discharge (case study 3.8). Connecticut used these relationships to develop a TMDL for a small stream identified as impaired based on biological assessments. Because the cause of impairment was unknown, an SI was completed. The SI determined that the most probable cause of impairment was the complex array of pollutants transported by stormwater into the stream. The TMDL is expressed as a reduction target for specific segments of the stream and is to be implemented through reduction of IC where practical and improved stormwater management throughout the watershed. Connecticut will evaluate progress toward the TMDL's implementation using biological assessments in conjunction with surface water chemistry assessments.

Additionally, EPA is encouraging states and tribes to develop TMDLs on a watershed basis (e.g., to bundle TMDLs together) to enhance program efficiencies and foster more holistic analysis. Ideally, TMDLs would be incorporated into comprehensive watershed strategies, while biological assessments would provide information on how the aquatic community responds to the full array of restoration activities. EPA is launching the Recovery Potential Screening Tools and Resources website (USEPA 2012),[9] designed to help state, tribal, and other restoration programs evaluate the relative restorability of impaired waters and help prioritize TMDL development. The website provides an approach to identify the use impaired waters and watersheds most likely to respond well to restoration, as well as information on methods, tools, technical information, and instructional examples that managers can customize for restoration programs in any geographic locality. Application of a gradient of biological response to levels of stress, like the Biological Condition Gradient (BCG) (see Section 2, Tool #

2, *The Biological Condition Gradient*), can provide a framework to help assess incremental progress in restoring a waterbody's aquatic life use and report environmental outcomes.

National Pollutant Discharge Elimination System Permits

Under section 402 of the CWA, point source discharges of pollutants to waters of the United States are covered by National Pollutant Discharge Elimination System (NPDES) permits. Under EPA regulations at 40 CFR 122.44(d), an NPDES permit must contain water quality-based effluents if it is found that a discharge will cause, have the reasonable potential to cause, or contribute to an excursion above a WQS. States must assess permitted effluent discharges in a manner that is consistent with EPA NPDES regulations (40 CFR 122.44).[10] States and tribes can use biological assessment information in addition to chemical-specific and WET data to support development of permit conditions that will protect water quality, including attainment of state WQS. Data from biological assessments can be used independently from, or in combination with, WET or chemical data to assess WQS attainment (USEPA 1991b). If any one or a combination of these three assessment methods demonstrates that the applicable WQS are not attained, appropriate and corrective action would be taken to address the findings as necessary, including compliance with applicable NPDES permit development provisions at 40 CFR PART 122.44(d)(1).

While narrative biological criteria might exist for many states and some authorized tribes in their WQS, in order for biological assessment information to effectively support the NPDES permit process there should be an EPA-approved numeric interpretation of the narrative biological criteria. States and tribes that have adopted biological criteria in their WQS may benefit from the use of biological assessment data as an additional biological check of permit controls, including limits, to see if they result in abating pollutant impacts, restoring water quality, or preventing further degradation. In addition, biological assessments as a "special studies/additional monitoring" permit condition can be used to assess overall permit effectiveness to control source pollutant(s) and used as an NPDES permit trigger to reopen and potentially modify the permit[11] if the biological assessment studies indicate that the permitted discharge continues to impact the receiving waterbody.

Also, while biological assessments can establish that aquatic life use impairment exists in the area of the discharge, the cause of the impairment might be wholly or partially due to point sources or NPS pollution. In such cases, an NPDES permit could establish controls based on the portion of impairment that is related to the effluent. Thus, additional chemical analysis and WET tests and/or source identification are typically conducted. For example, Vermont has used biological assessment information to support changes to effluent limits for metals on the basis of impact analysis, WET tests, and documented stressor-response relationships between metals and the aquatic biota (case study 3.9). That information helped support requiring additional treatment technologies that resulted in improved water quality. Upstream and downstream biological assessments were part of the follow-up monitoring plan and, with chemical and WET data, documented the resulting improvements in ambient biological and chemical conditions. Thus, in conjunction with required NPDES effluent monitoring such as WET and chemical-specific information, Vermont used biological assessments and its EPA-approved biological criteria to support narrative NPDES permit requirements to protect aquatic life. Currently Vermont has refined aquatic life uses (e.g., tiered aquatic life uses) and narrative biological criteria in its WQS supported by published peer-reviewed technical procedures for translating the narrative biological criteria into a numeric threshold.

NPS Pollution

Biological assessments can be a sensitive indicator of cumulative effects from multiple and unpredictable stressors from NPS pollution. Tracking water quality conditions using biological assessments is one way to assess whether the biological community is affected by NPS pollution and that efforts to improve degraded waters using voluntary BMPs are effective. In managing NPS pollution, a natural resource agency could initiate cooperative land use programs in an area or install BMPs to improve the water resource and establish biological goals as a benchmark for restoration. Before-and-after biological assessments compared to the biological benchmark make it possible to evaluate the success of management actions. For example, Michigan has used biological assessments to help determine biological impairments, target restoration efforts, and monitor results in Carrier Creek (case study 3.11).

Compliance Evaluation and Enforcement Support

Regulatory authorities can use biological assessment information to support enforcement actions by helping to document biological impacts and measure recovery of the aquatic community due to mitigation and cleanup actions. For example, a fish kill in a tributary to the Potomac River in Maryland and the District of Columbia was caused by illegal dumping of pesticide wastes in Maryland. Biological and chemical sampling data were used to locate the source of the pesticide wastes, identify the responsible party, and show subsequent improvements in water quality as a result of enforcement activities (case study 3.12). Biological assessment information, in conjunction with biological assays and chemical and physical assessments, can assist enforcement agencies in assessing damage and levying fair and reasonable damage assessments on those proven responsible for toxic spills, and determining the rate and level of stream recovery.

Watershed Protection

Increasingly, EPA, states, territories, and tribes are implementing CWA programs on an integrated watershed basis—including air, land, and ecosystem relationships and related regulatory tools such as those used in the Chesapeake Bay[12] and the National Estuary programs (NEPs)[13] (USEPA 2007). Biological assessments are used in watershed-level programs to help define ecological goals and assess progress in achieving those goals. Recently, EPA has embarked on the Healthy Watershed Initiative, which focuses on protecting high-quality waters and watersheds (USEPA In draft). It is a strategic approach that identifies healthy waters and watersheds at the state level and then targets resources at both the state and local levels for their protection. Biological assessments provide critical information and measurable benchmarks to identify high-quality waters in healthy watersheds and then, over time, evaluate how effectively such systems are being protected. The State of Virginia is using biological assessments in its own Healthy Watersheds initiative to define protection and restoration goals that resonate with the public (case study 3.14). EPA's Office of Research and Development (ORD) is working with several states, territories, and NEPs to develop biological assessment tools and approaches that can be applied at multiple scales to protect estuarine and coastal ecosystems and their watersheds (case study 3.16). Additionally, the BCG (see Section 2, Tool # 2) can be applied as a field-based assessment framework to describe the health of waterbodies and their watersheds and communicate the biological condition to the public (USEPA In draft). And, in conjunction with refined aquatic life uses and biological criteria adopted into WQS, a BCG-like framework can be used to support management actions to protect existing high-quality waters in a healthy watershed, as demonstrated by the State of Maine (case study 3.1).

SECTION 2. TOOLS FOR IMPROVING THE USE OF BIOLOGICAL ASSESSMENTS IN WATER QUALITY MANAGEMENT

EPA has published several documents that provide guidance on incorporating biological assessment information into water quality programs, many of which have been in use for several years. They include technical guidance on developing biological criteria and general program guidance on application of biological assessment information in different water quality programs. A summary of these documents is provided in Appendix A. Additionally, other technical support documents, or technical tools, have been recently developed to further assist states and tribes in developing robust biological assessment programs and applying biological assessment information. Three of these recent tools are listed below and briefly summarized in the following pages.

- ***Tool #1: The Biological Assessment Program Review.*** The level of program rigor determines how well the monitoring and assessment program produces the information needed to support management decision making. A review process and checklist have been developed and piloted by regions, states, and tribes to help assess the technical capability of a state or tribal biological assessment program and strategically determine where to invest resources to develop a technically robust biological assessment program.
- ***Tool #2: The Biological Condition Gradient (BCG).*** The BCG is a conceptual model that describes how biological attributes of aquatic ecosystems might change along a gradient of increasing anthropogenic stress. The model can serve as a template for organizing field data (biological, chemical, physical, landscape) at an ecoregional, basin, watershed, or stream segment level. A BCG calibrated with field data can help states and tribes more precisely define biological expectations for their designated aquatic life uses, interpret current condition relative to CWA objective and goals, track biological community response to management actions, and communicate environmental outcomes to the public. The BCG was designed to help map different biological indicators on a common scale of biological condition to facilitate communication among

programs and across jurisdictional boundaries. The BCG is currently being field tested in several regions and states.

- ***Tool #3: Stressor Identification (SI) and Causal Analysis/Diagnosis Decision Information System (CADDIS).*** In 2010 EPA updated its technical support document on causal analysis and literature database to help states and tribes identify the most probable cause of impairment to a waterbody. Specific databases on biological response to stress have been compiled and will undergo continuous updating so that the best available and peer-reviewed literature will be accessible as part of CADDIS. This document and database will assist states that have listed waters as impaired on the basis of biological assessments when the cause of impairment is not known.

2.1. Tool #1: Biological Assessment Program Review

Purpose:

To provide a stepwise process to assist states in evaluating the technical capability of their biological assessment programs and to strategically determine where to invest resources to enhance the technical capability of their programs.

This tool can be used to answer questions, including the following:

- Does the quality of data being generated support the management decisions I need to make?
- What are the strengths and needs of my existing program?
- How do I build on my current program and further strengthen it?

Source: EPA's website on key concepts for using biological indicators: http://www.epa.gov/bioiweb1/html/key concepts.html.

The information provided below describes technical elements of a biological assessment program, summarizes the process and benefits of conducting a program review, and discusses regional/state pilot programs.

The Program Review Process

The critical technical elements review is a systematic ***process*** to evaluate biological assessment program rigor and to identify logical next steps for overall program improvement.

The document provides a ***template*** for evaluating critical technical components of a biological assessment program that are scored to arrive at a level of program rigor, from level 1 (the least rigorous) to level 4 (the most rigorous) (Table 2-1).

The review provides a framework for identifying programmatic strengths and weaknesses and helps program managers and technical staff members determine key tasks to upgrade the technical abilities of their program (Figure 2-1). The evaluation process also identifies opportunities to improve integration of WQS and monitoring and assessment programs. This review process was initially piloted in EPA Region 5 and more recently applied and further refined in Region 1. Initial programs reviews have focused on biological assessments of streams and rivers, but with some refinements in methodology this evaluation process can be applied to other types of waterbodies. The states have used the results of the review to target resources and prioritize actions to strengthen the technical basis of their biological assessment programs.

The first part of the review involves discussion on the design of the existing monitoring and assessment program, the degree to which there is systematic collection of data from the environment, and how well the data analysis produces information suitable for making the various decisions asked of it—such as determining attainment of aquatic life uses, identifying high-quality waters for antidegradation purposes on a waterbody-by-waterbody basis, evaluating the severity and extent of impairments, and supporting causal analysis and pollutant source identification (i.e., toxicity identification evaluation [TIE] and toxicity reduction evaluation [TRE]). It is essential that experts in the different program areas be engaged in the discussions to help ensure that data quality and information requirements are accurately represented and properly implemented, especially with regard to EPA-published methodologies.

The information helps document how monitoring and assessment information is used to support the reporting requirements mandated by the CWA and other state or tribal efforts to characterize the status of waterbodies and plan for implementing restoration efforts. This part of the program review might also examine how the state or tribe uses biological assessment information to more precisely define aquatic life uses and develop biological criteria.

Table 2-1. Key features of the technical attributes for levels of rigor in state/tribal biological assessment programs (streams and rivers). (Terms in the table are included in the glossary, this template can be modified and applied to other waterbody types)

	Attributes of levels of biological assessment program rigor			
Key features	Level 1	Level 2	Level 3	Level 4
Temporal and spatial coverage	Variable data collection times; upstream/downstream and fixed stations	Index period for convenience; non-random design at a coarse scale (e.g., 4- to 8-digit hydrologic unit code [HUC])	Calibrated seasonal index periods; statewide spatial design using rotating basins at a coarse scale (e.g., 4- to 8-digit HUC)	Scientifically-derived temporal sampling for management decisions; multiple spatial designs for multiple issues; 11- to 14-digit HUC
Natural classification of aquatic ecosystems	No partitioning of natural variability; no incorporation of differences in stream characteristics such as size, gradient	Classification usually a geo-graphical or other similar organization (e.g., fishery-based cold or warmwater; lacks intra-regional strata [size, gradient])	Classification based on a combination of landscape features and physical habitat structure; considers all intra-regional strata and specific ecosystems	Fully partitioned and stratified classification scheme that transcends jurisdictions and recognizes zoogeographical aspects of assemblages
Reference conditions	No reference conditions; presence and absence of key taxa are based on best professional judgment	A site-specific control or paired watershed approach can be used for assessment; regional reference sites are lacking	Reference conditions used in watershed assessments; regional reference sites are too few in number or spatial density	Regional reference conditions are established in the applicable waterbody ecotypes and aquatic resource classes
Sampling and sample processing	Approach is cursory and relies on operator skill and best professional judgment, producing highly variable and less comparable results	Textbook methods are used rather than in-house development of standard operating procedures to specify methods	Methods are calibrated for state purposes and are detailed and well documented; supported by in-house testing and development	Same as Level 3, but methods cover multiple assemblages; high taxonomic resolution
Data management	Sampling event data are organized in a series of spreadsheets	Separate databases are used for physical, chemical, and biological data with separate GIS shapefiles of sites	A true relational database is specifically designed to include data validation checks (e.g., Oracle, SQL Server, Access)	Relational database of biological assessment data with automated data review validation tools and geospatial analysis

Key features	Attributes of levels of biological assessment program rigor			
	Level 1	Level 2	Level 3	Level 4
Biological endpoints and thresholds	Assessment based on presence or absence of targeted or key species; attainment thresholds are not specified and no BCG	A biological index or endpoint is by specific waterbodies; single dimension measures used	A biological index, or model, developed and calibrated for use throughout the state for the various waterbody types	Biological indexes, or models, for multiple assemblages are developed and calibrated for a state and uses the BCG
Causal analysis	Support for causal analysis is lacking	Coarse indications of response via assemblage attributes at gross level (i.e., general indicator groups)	Developed indicator guilds and other aggregations to support causal associations; diagnostic capability is supported by studies	Response patterns are most fully developed and supported by extensive research and case studies across spatial and temporal scales

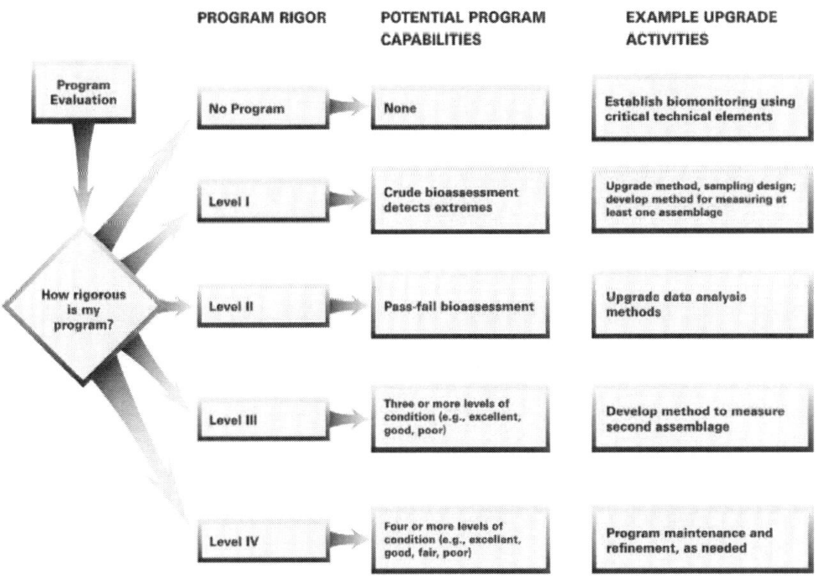

Figure 2-1. Key features of the program review process and examples of commensurate upgrades.

Evaluation of Critical Technical Elements of a State's or Tribe's Biological Assessment Program

The program review evaluates 13 critical technical elements of a biological assessment program associated with design, methods, and data interpretation (e.g., survey design, method of classification, procedures to establish reference conditions, protocols for sampling collection and processing, data management and analysis, formal peer review). On the basis of the discussions in the first phase of the review, where program information needs are identified, a list of recommendations is developed according to the strengths and gaps identified in the technical program evaluation. The recommendations are presented in a logical, stepwise progression so that a state or tribe can build on its technical program strengths and target resources effectively to address the program gaps. Participation of program managers and technical staff representing different water quality programs is important in the review to build a shared understanding and broad perspective on existing use of biological assessment information and begin to identify the technical program gaps and areas for improved use.

Case Example: Technical Evaluations in Minnesota and Connecticut

The Minnesota Pollution Control Agency (MPCA) decided in 2005 to use biological assessment information to develop refined aquatic life uses and numeric biological criteria in its WQS to meet its objectives of setting management goals for waterbodies on the basis of their best potential condition. MPCA also found biological assessment information as useful to educate and engage stakeholders and the public. MPCA used the Critical Technical Elements Program Evaluation process to determine where its program was in 2005 and what tasks were yet to be accomplished to reach its stated goals. Using the findings, MPCA developed a detailed plan for developing a technical program sufficiently rigorous to support adoption in the state's WQS in 2011–2014 of the most appropriate aquatic life uses and numeric biological criteria. MPCA continues to follow the plan, addressing the priority recommendations identified in the program evaluation, and is proceeding with biological criteria development. As part of this effort, MPCA is exploring application of the BCG, the second tool discussed in this document, to develop biological goals for their waters that are tailored to specific waterbody types and uses.

The Connecticut Department of Environmental Protection (CT DEP) has been monitoring aquatic biological conditions using benthic macroinvertebrates since the late 1980s and has steadily upgraded its technical program over the years. The state operates a statewide monitoring and assessment program that includes multiple spatial designs to produce both statewide assessments using probabilistic design and listings of impaired waters using targeted sampling design. CT DEP underwent a Critical Elements Program Evaluation in 2006 to help identify and prioritize additional technical program improvements needed to develop numeric biological criteria for different levels of quality along a gradient of condition (e.g., excellent and good quality waters). The program was evaluated at a level 2 with specific tasks identified to build its technical capability (e.g., improved spatial resolution in watershed assessment design from 8-digit HUC to 10- to 12-digit HUC; a regionally-calibrated multimetric index for benthic macroinvertebrates and one for fish that distinguishes between coldwater and warmwater assemblages; instituting an independent peer review process). Since the review, CT DEP has improved the technical capability of the biological assessment program to a level 3 and now has two numeric indices and enhanced spatial monitoring design.

These examples show how states and tribes can use the results of the Critical Elements Program Evaluation to develop a blueprint for making orderly improvements and attaining the technical proficiency to respond to management questions and improve decision making—including support for condition assessments, attainment of WQS, diagnosis of biological impairment, and effectiveness monitoring. The program review process identifies specific and successive improvements that are needed to improve the rigor of the biological assessment program and a checklist so that progress can be identified and tracked.

2.2. Tool #2: The Biological Condition Gradient

Purpose: To provide a common scale of biological condition to support comparisons between programs and across jurisdictional boundaries.

This tool can be used to help answer questions, including the following:

- What biological community should be at a site, e.g., natural conditions?

- Are we protecting our high-quality waters?
- Are we making progress to restore our degraded systems?
- Are our actions making real and lasting environmental improvements?

Source: The biological condition gradient: A descriptive model for interpreting change in aquatic ecosystems (Davies and Jackson 2006).

This section provides an overview of the BCG and how it can be calibrated for specific use by a state or tribe. The BCG is being applied and tested in several regions and states.

What Is the BCG?

Over the past 40 years, states have independently developed technical approaches to assess biological condition and set designated aquatic life uses for their waters. The BCG was designed to provide a means to map different indicators on a common scale of biological condition to facilitate comparisons between programs and across jurisdictional boundaries in context of the CWA. The BCG is a conceptual, narrative model that describes how biological attributes of aquatic ecosystems change along a gradient of increasing anthropogenic stress. It provides a framework for understanding current conditions relative to natural, undisturbed conditions (Figure 2-2). Some states, such as Maine and Ohio, have used a framework similar to the BCG to more precisely define their designated aquatic life uses (case studies 3.1 and 3.5).

Agreeing that, even in different geographic and climatological areas, a similar sequence of biological alterations occurs in streams and rivers in response to increasing stress, biologists from across the United States developed the model (Davies and Jackson 2006). The model shows an ecologically based relationship between the stressors affecting a waterbody (e.g., physical, chemical, biological impacts) and the response of the aquatic community (i.e., biological condition). The model is consistent with ecological theory and can be adapted or calibrated to reflect specific geographic regions and waterbody type (e.g., streams, rivers, wetlands, estuaries, lakes). Approaches to calibrate the BCG to region-, state-, or tribe-specific conditions are being piloted in several ecological regions by multiple states and tribes.

In practice, the BCG is used to first identify the critical attributes of an aquatic community (see Table 2-2) and then describe how each attribute changes in response to stress. Practitioners can use the BCG to interpret biological condition along a standardized gradient, regardless of assessment method, and apply that information to different state or tribal programs. For example, Pennsylvania is exploring the use of a BCG calibrated to its streams to complement its existing biological indices for macroinvertebrates and to describe the biological characteristics of waters along a gradient of condition. The state is evaluating using this information to help assess aquatic life use impairments and to describe waters of the highest quality (case study 3.4).

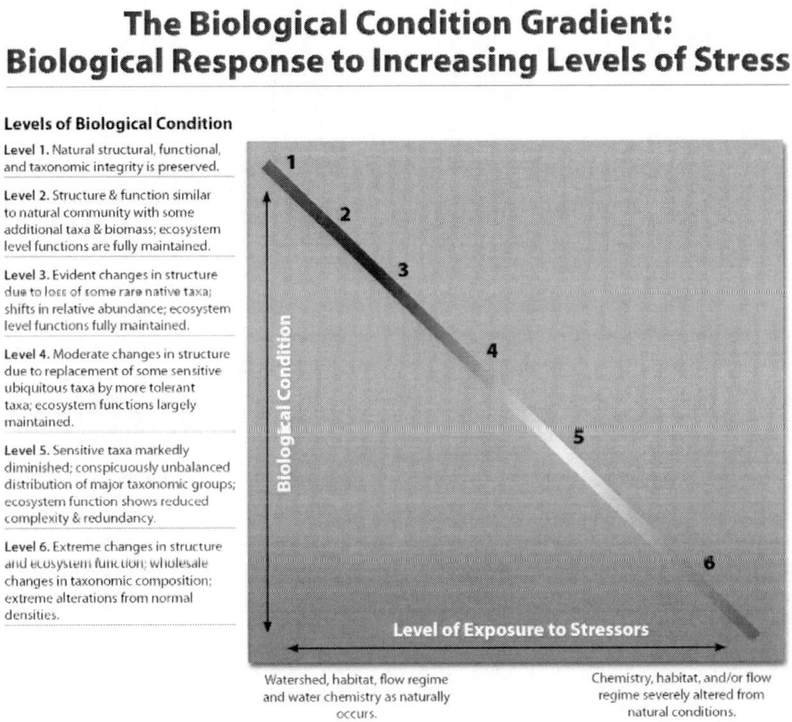

Source: Modified from Davies and Jackson 2006.

Note: The BCG was developed to serve as a scientific framework to synthesize expert knowledge with empirical observations and develop testable hypotheses on the response of aquatic biota to increasing levels of stress. It is intended to help support more consistent interpretations of the response of aquatic biota to stressors and to clearly communicate this information to the public, and it is being evaluated and piloted in several regions and states.

Figure 2-2. The BCG.

The BCG model provides a framework to help water quality managers do the following:

- Decide what environmental conditions are desired (goal-setting)—The BCG can provide a framework for organizing data and information and for setting achievable goals for waterbodies relative to "natural" conditions (e.g., condition comparable or close to undisturbed or minimally disturbed condition).
- Interpret the environmental conditions that exist (monitoring and assessment)—Practitioners can get a more accurate picture of current waterbody conditions.
- Plan for how to achieve the desired conditions and measure effectiveness of restoration—The BCG framework offers water program managers a way to help evaluate the effects of stressors on a waterbody, select management measures by which to alleviate those stresses, and measure the effectiveness of management actions.
- Communicate with stakeholders—When biological and stress information is presented in this framework, it is easier for the public to understand the status of the aquatic resources relative to what high-quality places exist and what might have been lost.

How Is the BCG Constructed?

The BCG is divided into six levels of biological conditions along the stressor-response curve, ranging from observable biological conditions found at no or low levels of stress (level 1) to those found at high levels of stress (level 6) (Figure 2-2). The technical document provides a detailed description of how 10 attributes of aquatic ecosystems change in response to increasing levels of stressors along the gradient, from level 1 to 6 (see Table 2-2). The attributes include several aspects of community structure, organism condition, ecosystem function, spatial and temporal attributes of stream size, and connectivity.

Each attribute provides some information about the biological condition of a waterbody. Combined into a model like the BCG, the attributes can offer a more complete picture about current waterbody conditions and also provide a basis for comparison with naturally expected waterbody conditions. All states and tribes that have applied a BCG used the first seven attributes that describe the composition and structure of biotic community on the basis of the tolerance of species to stressors and, where available, included information on the presence or absence of native and nonnative species and, for fish and amphibians, observations on overall condition (e.g., size, weight, abnormalities, tumors).

The last three BCG attributes of ecosystem function and connectance and spatial and temporal extent of detrimental effects can provide valuable information when evaluating the potential for a waterbody to be protected or restored. For example, a manager can choose to target resources and restoration activities to a stream where there is limited spatial extent of stressors or there are adjacent intact wetlands and stream buffers or intact hydrology versus a stream with comparable biological condition but where adjacent wetlands have been recently eliminated, hydrology is being altered, and stressor input is predicted to increase. Pennsylvania is evaluating indicators comparable to the BCG spatial and connectance attributes IX and X to characterize the biological conditions of streams in healthy watersheds where resources may be well spent to successfully protect such waters (see case study 3.4). Additionally, several of EPA's NEPs, in conjunction with EPA ORD, are exploring application of those attributes at a whole-estuary scale (e.g., distribution and connectance of critical aquatic habitats and associated biota) (see case study 3.16).

Additionally, individual attributes might uniquely respond to a specific stressor or group of associated stressors (biological response signatures) (Yoder and Rankin 1995; Yoder and Deshon 2003). That information could contribute to the causal analysis of biological impairment discussed in Tool #3, *Stressor Identification (SI) and Causal Analysis/Diagnosis Decision Information System (CADDIS).*

Table 2-2. Biological and other ecological attributes used to characterize the BCG

Attribute	Description
I. Historically documented, sensitive, long-lived, or regionally endemic taxa	Taxa known to have been supported according to historical, museum, or archeological records, or taxa with restricted distribution (occurring only in a locale as opposed to a region), often due to unique life history requirements (e.g., sturgeon, American eel, pupfish, unionid mussel species).
II. Highly sensitive (typically uncommon) taxa	Taxa that are highly sensitive to pollution or anthropogenic disturbance. Tend to occur in low numbers, and many taxa are specialists for habitats and food type. These are the first to disappear with disturbance or pollution (e.g., most stoneflies, brook trout [in the east], brook lamprey).

Table 2-2. (Continued)

Attribute	Description
III. Intermediate sensitive and common taxa	Common taxa that are ubiquitous and abundant in relatively undisturbed conditions but are sensitive to anthropogenic disturbance/pollution. They have a broader range of tolerance than attribute II taxa and can be found at reduced density and richness in moderately disturbed sites (e.g., many mayflies, many darter fish species).
IV. Taxa of intermediate tolerance	Ubiquitous and common taxa that can be found under almost any conditions, from undisturbed to highly stressed sites. They are broadly tolerant but often decline under extreme conditions (e.g., filter-feeding caddisflies, many midges, many minnow species).
V. Highly tolerant taxa	Taxa that typically are uncommon and of low abundance in undisturbed conditions but that increase in abundance in disturbed sites. Opportunistic species able to exploit resources in disturbed sites. These are the last survivors (e.g., tubificid worms, black bullhead).
VI. Nonnative or intentionally introduced species	Any species not native to the ecosystem (e.g., Asiatic clam, zebra mussel, carp, European brown trout). Additionally, there are many fish native to one part of North America that have been introduced elsewhere.
VII. Organism condition	Anomalies of the organisms; indicators of individual health (e.g., deformities, lesions, tumors).
VIII. Ecosystem function	Processes performed by ecosystems, including primary and secondary production; respiration; nutrient cycling; decomposition; their proportion/dominance; and what components of the system carry the dominant functions. For example, shift of lakes and estuaries to phytoplankton production and microbial decomposition under disturbance and eutrophication.
IX. Spatial and temporal extent of detrimental effects	The spatial and temporal extent of cumulative adverse effects of stressors; for example, groundwater pumping in Kansas resulting in change in fish composition from fluvial dependent to sunfish.
X. Ecosystem connectance	Access or linkage (in space/time) to materials, locations, and conditions required for maintenance of interacting populations of aquatic life; the opposite of fragmentation. For example, levees restrict connections between flowing water and floodplain nutrient sinks (disrupt function); dams impede fish migration, spawning.

Source: Modified from Davies and Jackson 2006.

Calibrating the Conceptual Model to Local Conditions

The BCG can serve as a starting point for defining the response of aquatic biota to increasing levels of stress in a specific region. Although the BCG was developed primarily using forested stream ecosystems, the model can be applied to any region or waterbody by calibrating it to local conditions using specific expertise and local data. To date, most states and tribes are calibrating the BCG using the first seven attributes that characterize the biotic community primarily on the basis of tolerance to stressors, presence/absence of native and nonnative species, and organism condition. Although the model has been developed for six levels of condition, six levels might not be necessary or feasible depending on limitations in data or level of technical rigor (see Section 2, Tool #1, *Biological Assessment Program Evaluation*) or naturally occurring conditions. For example, ephemeral streams in the arid Southwest naturally support a community of aquatic organisms that tolerate extreme conditions that range from intense, monsoon-like precipitation to extensive periods of drought. Those organisms might also be able to tolerate the presence of stressors. Thus, the range of response to anthropogenic stress in such streams (e.g., moderately tolerant to very tolerant species) might be abbreviated compared to that of a forested stream community in a temperate climate (e.g., very sensitive to very tolerant species). Three or four tiers might be suitable for those waters.

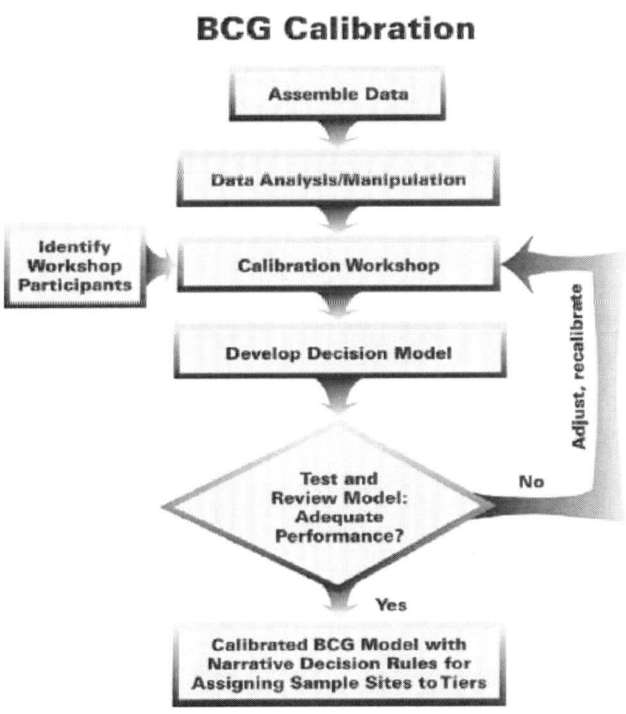

Figure 2-3. Steps in a BCG calibration.

It is a multistep process to calibrate a BCG to local conditions (Figure 2-3). That process is followed to describe the native aquatic assemblages under natural conditions; identify the predominant regional stressors; and describe the BCG, including the theoretical foundation and observed assemblage response to stressors. Calibration begins with the assembly and analysis of biological monitoring data. Next, a calibration workshop is held in which experts familiar with local conditions use the data to define the ecological attributes and set narrative statements. For example, narrative decision rules for assigning sites to a BCG level on the basis of the biological information collected at sites. New Jersey is one of several states that are field testing this approach. Documentation of expert opinion in assigning sites to tiers is a critical part of the process. A decision model can then be developed that encompasses those rules and is tested with independent data sets. A decision model based on the tested decision rules is a transparent, formal, and testable method for documenting and validating expert knowledge (see Table 2-3 for examples). A quantitative data analysis program can then be developed using those rules. EPA recommends peer review of model.

Table 2-3. Example of narrative decision rules for distinguishing BCG Level 2 from Level 3 for streams, modified from New Jersey BCG expert workshop

Attributes	Rules for BCG Level 2 Structure and function of community similar to natural community with some additional taxa and biomass	Rules for BCG Level 3 Evident changes in structure due to loss of some rare native taxa; shifts in relative abundance
Total taxa	More than 12 taxa	More than 12 taxa
Highly Sensitive Taxa (Attribute II only)	More than two taxa	May be absent
Richness of Sensitive Taxa (combination of Attributes II and III,see table 2-2)	Attribute II + Attribute III are more than 50% of total taxa richness	Attribute II + Attribute III are more than 35% of total taxa richness
Abundance of Tolerant Taxa (Attribute V)	Abundance of Attribute V is less than 20% of community	Abundance of Attribute V is less than 50% of community

In the example above, both BCG levels 2 and 3 support comparable levels of overall taxa (e.g., total taxa). However, there is a shift from BCG level 2 to BCG level 3 in proportion and abundance of sensitive and tolerant taxa (e.g., a decrease in proportion of sensitive taxa and an increase in abundance of pollution-tolerant taxa). The BCG describes incremental shifts in community composition and other biological parameters along a gradient of increasing anthropogenic stress. The BCG can be used to detect measurable changes in the aquatic biota before there is a complete loss of a certain type or category of taxa such as loss of pollution sensitive or native species. This tool will enable earlier detection and support action to prevent loss of species or other biological changes. This tool can be used to raise the discriminatory power of biological assessment programs in a nationally consistent, transparent manner. Narrative decision rules are the first step in formalizing expert opinion and expressing empirical findings that can then be tested and validated.

Case Example: New Jersey BCG Calibration

New Jersey developed and calibrated a BCG for its upland streams. The New Jersey Department of Environmental Protection (NJ DEP) convened an expert panel workshop that included aquatic biologists and water quality experts familiar with the aquatic fauna that inhabit these streams. The panel developed descriptions of the ecological attributes for these streams in New Jersey and created the narrative rules for assigning sites to levels along the stressor gradient.

The expert panel reviewed the list of taxa from the New Jersey Ambient Biological Monitoring Network to assign taxa to attributes I–VI. Next, the panel examined macroinvertebrate data from 58 upland stream sites and reached consensus on the level assignments for all sites reviewed. The panel was able to distinguish five separate levels (levels 2–6, see below) for New Jersey upland streams. The first level described in Davies and Jackson (2006) consists of entirely pristine sites and was not included because the panel could not identify any level 1 (pristine) sites in New Jersey. On the basis of the characterization of sites identified as belonging to different BCG levels, the panel developed a set of narrative decision rules and descriptions for distinguishing among the levels.

BCG level 2 (Minimal changes in structure and function)—Because of extensive historical land clearing, cultivation, and early industrial use followed by abandonment and reforestation from the early 20th century, the least stressed watersheds are thought to reflect at best BCG level 2.

Most of the 19th century legacy is in changed stream morphology and hydrology that persist in valley bottoms (Walter and Merritts 2008). Watersheds are predominantly forested, with recreational use but little residential or agricultural use. The group consensus was that several richness criteria (i.e., total taxa, highly sensitive taxa, and all sensitive taxa) must all be met for a site to be considered to be in level 2.

BCG level 3 (Evident changes in structure and function)—A typical level 3 stream has a largely forested watershed but some areas of suburban development or limited agriculture. Criteria for level 3 are similar to those for level 2, but richness of the sensitive organisms is somewhat reduced and sensitive organisms do not numerically dominate the assemblage. All the criteria for level 3 were considered critical.

BCG level 4 (Moderate changes in structure and function)—Typical level 4 streams in New Jersey often have relatively extensive suburban and commercial development, some agricultural land use, but substantial areas of natural land cover, often mixed with residential areas. In BCG level 4, the sensitive taxa are present and still constitute a significant fraction of the community, but they are far reduced below their dominance in level 2 and their subdominance in level 3. The assemblage has degraded but maintains ecosystem functions as represented by the sensitive taxa.

BCG level 5 (Major changes in structure and function)—BCG level 5 is discriminated from level 4 by a significant reduction of sensitive taxa (attributes II and III) to the point where they are merely incidental if present and are not a functional part of the community. Although BCG level 5 can have high abundance and high taxa richness, the assemblage is dominated by intermediate and tolerant taxa, and sensitive taxa have all but disappeared.

BCG level 6 (Severe changes in structure and function)—BCG level 6 reflects nearly complete disruption and degradation of the biological community to either very low abundance (less than 50 organisms in New Jersey's standard sampling procedure) or very low taxon richness. While extremely low abundance often indicates toxic conditions, extremely low richness coupled with high abundance often indicates organic enrichment and high-density urban runoff.

New Jersey is considering using the calibrated BCG and the narrative decision rules to help identify high-quality waters on a waterbody-by-waterbody basis for antidegradation purposes.

2.3. Tool #3: Stressor Identification (SI) and Causal Analysis/Diagnosis Decision Information System (CADDIS)

> Purpose: To identify the cause of aquatic life impairment when a waterbody is listed because of biological impairment and the cause is unknown.
>
> This tool can be used to answer questions such as the following:
>
> - How can I use biological and stressor information to identify cause of biological impairment?
>
> Sources: *Stressor Identification Guidance Document* (USEPA 2000a); EPA's CADDIS website:
> http://www.epa.gov/caddis

This section describes how biological assessment information can be used to help identify stressors for impaired waters where cause of impairment is unknown.

How Can Biological Information Be Used for Stressor Identification?

Once a biological impairment has been determined, water quality managers examine existing water quality and landscape data and information to determine the cause and source of impairment, also known as stressor identification (SI). Typically, states and tribes identify the probable causes of the impairment and then, step-by-step, implement additional controls or management practices (or both) to fix the problem. Monitoring the response of the biota to management actions then helps to provide the necessary information on whether the primary stressors were correctly identified and the management actions effective. The biological response information provided in the initial assessment often includes useful information for identifying stressors; for example, the relationship between biological indicators and stressors such as the disappearance of certain benthic species sensitive to a specific toxin (e.g., sensitivity of aquatic life stage of mayflies to metal toxicity) or a shift in dominate community traits related to the increase of a stressor (e.g., a change in primary producer base because of zebra mussel invasion). Additionally, states and tribes have successfully implemented management actions that address co-occurring stressors supported by documented improvements in water quality. Maryland and the District of Columbia were able to use biological assessment data to document the biological effects of a pesticide spill that resulted in a fish kill in Rock Creek, a tributary to the Potomac River. The information was used as the basis for enforcement actions, and subsequent data were able to support a quantitative assessment of the biological impact and evidence of stream recovery (case study 3.12).

Stressor ID/CADDIS

In 2000 EPA's Office of Water and ORD developed a process for identifying any type of stressor or combination of stressors that causes biological impairment. The *Stressor Identification Guidance Document* (USEPA 2000a) is intended to lead water resource managers through a formal and rigorous process that identifies stressors causing biological impairment in aquatic ecosystems and provides a structure for organizing the scientific evidence supporting the conclusions.

The SI process is prompted by biological assessment data indicating that a biological impairment has occurred. The general SI process entails critically reviewing available information, forming possible stressor scenarios that might explain the impairment, analyzing those scenarios, and providing conclusions about which stressor(s) are causing the impairment. The SI process is iterative, usually beginning with a retrospective analysis of available data. The accuracy of the identification depends on the quality of data and other information used in the SI process. In some cases, additional data collection might be necessary to accurately identify the stressor(s). The conclusions can be translated into management actions, and the effectiveness of those management actions can be monitored (Figure 2-4).

The core of the SI process consists of the following three main steps:

- Listing candidate causes of impairment.
- Analyzing new and previously existing data to generate evidence for each candidate cause.
- Producing a causal characterization using the evidence generated to draw conclusions about the stressors that are most likely to have caused the impairment.

Again, the SI process is iterative. Practitioners will begin by analyzing available data to see if sufficient information is already available. The kinds of information needed include information on the type of impairment, the extent of the impairment, any evidence of the usual causes of impairment (e.g., hydrologic alteration, invasive species, habitat loss, toxicants, total nitrogen and phosphorus), and other information from the site. The evidence is considered first and then other, less direct kinds of evidence are gathered and evaluated, if needed. For example, one might consider other situations that are similar and can provide useful insights.

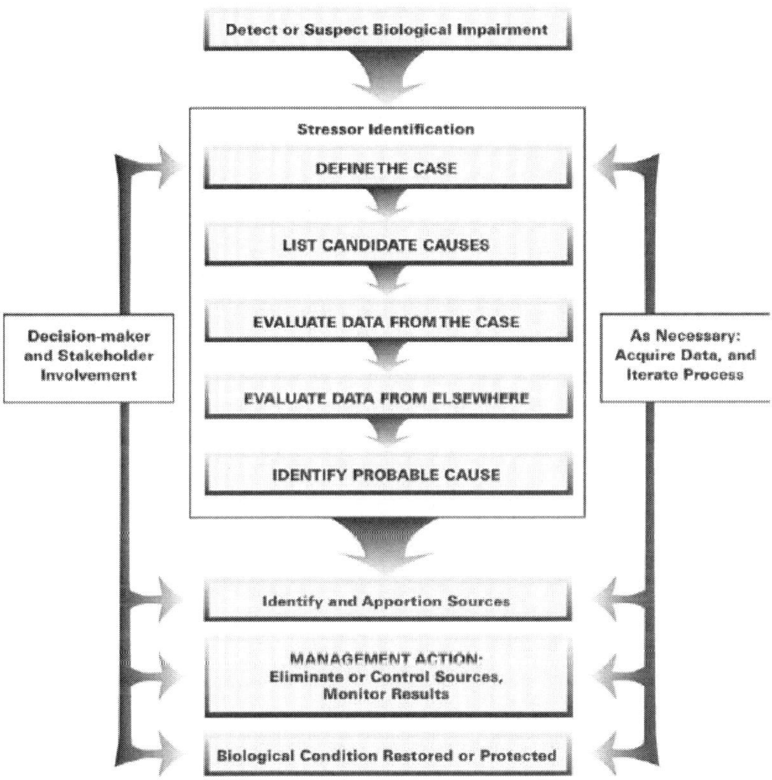

Source: USEPA 2010b.

Figure 2-4. Stressor identification process.

CADDIS is an online application of the SI process that uses a step-by-step guide, worksheets, technical information, and examples to help scientists and engineers find, access, organize, share, and use environmental information to evaluate causes of biological effects observed in aquatic systems such as streams, lakes, and estuaries.[14] CADDIS also contains updates, clarifications, and additional material developed since the SI guidance document was published in 2000.

Case Example: Nutrient Management in the Little Miami River, Ohio

In the early 1980s, Ohio EPA designated the Little Miami River as an Exceptional Warmwater Habitat (EWH) following the first complete biological survey of the mainstem and key tributaries in the Ohio WQS under the new system of tiered aquatic life uses adopted in 1978. While not all sites sampled in 1983 attained the EWH biological criteria for both the fish and macroinvertebrate assemblages, sufficient sites did attain the EWH use, thus demonstrating the potential for attainment of that use as long as critical habitat were present.

In 1988, more stringent effluent limits for typical wastewater treatment plant (WWTP) parameters (e.g., biochemical oxygen demand [BOD], ammonia-N, common heavy metals) were established for municipal WWTPs. In 1993, as part of the Ohio EPA rotating basin approach, both water quality and biological improvements were observed, accompanied by increase in waters achieving the EWH use. These improvements resulted from water quality-based permitting at municipal WWTPs and compliance with more stringent effluent limits. However, suburban development in the surrounding communities resulted in increased WWTP flows and loads through the 1990s and the level of stress on aquatic systems increased. In 1998 biological assessment results again documented a decline in EWH attainment. The decline was associated with increased phosphorus loadings, which had not been targeted as part of the earlier water quality-based permitting. Additionally, increased diel dissolved oxygen variations and elevated phosphorus concentrations were observed. Following a determination that the observed degradation was related to loadings discharged primarily during summer low flows (i.e., from municipal WWTPs), the largest WWTPs implemented a phased reduction of phosphorus loadings through NPDES permits. A follow-up biological assessment in 2007 documented attainment of the EWH biological criteria along most of the mainstem of the Little Miami River after point source phosphorus controls were implemented. The findings documented the effectiveness of the nutrient removal provided by the WWTPs and confirmed the original hypothesis that the biological impairments were indeed linked to phosphorus loadings discharged by the point sources.

This example highlights the value of conducting before-and-after biological assessments to support NPDES permitting.

Source: Ohio EPA (Environmental Protection Agency). 2009. Biological and Water Quality Study of the Lower Little Miami River and Selected Tributaries 2007 Including the Todd Fork Watershed. Watershed assessment units 05090202 06, 07, 08, 09 and 14. Clermont, Clinton, Hamilton, and Warren counties. Ohio EPA technical report EAS/2009-10-06. 201 pp.

Case Example: Causal Assessments of Impairment in Iowa

The Iowa Department of Natural Resources (IDNR) identified causes of biological impairment of the Little Floyd River using EPA's SI methodology (Haake et al. 2010). Through its biological monitoring program and using Iowa's benthic macroinvertebrate index, IDNR identified the Little Floyd River as impaired, with biotic index scores well below the reference population for the area. IDNR applied the SI process to biological, chemical, and physical data collected from the river.

Candidate causes for the biological impairment were flow alteration, substrate alteration, turbidity, altered basal food source, low dissolved oxygen concentrations, high temperature, and high ammonia concentrations. Biological metrics specific to the impairment were used to identify a less impaired location in the stream to help discover the cause of more severe effects in other parts of the stream. These paired biological, physical, and chemical data from the stream were used to develop evidence of co-occurrence of exposure and effects and evidence of preceding causation; that is, the presence of sources and mechanistic pathways leading to conditions where exposure could occur. Evidence that the exposure level was sufficient to cause either the fish or the invertebrate effects was developed from two Iowa data sets with paired biological, physical, and chemical data. The interquartile range of values for the various stressors from ecoregion reference sites were compared to the values observed for the Little Floyd River. Also, the mean value at statewide random sites was compared to the values in the Little Floyd River. All the supporting or discounting evidence was weighted, and the body of evidence for each candidate cause was weighed.

The formal process revealed that sediment deposition, hypoxia, heat stress, and ammonia toxicity were probable causes of impaired biological condition in the Little Floyd River.

Other causes were discounted if they were unlikely or deferred if the data were insufficient to make a determination. The assessment was used to develop a recovery plan for the stream and was a contributing impetus for developing temperature criteria as part of IDNR's WQS. Without Iowa's basic commitment to integrated monitoring and use of biological, physical, and chemical data, the analysis and the SI would not have been possible.

Source: Haake, D.M., T. Wilton, K. Krier, A.J. Stewart., and S.M. Cormier. 2010. Causal assessment of biological impairment in the Little Floyd River, Iowa, USA. Human Ecological Risk Assessment 16(1):116–148.

SECTION 3. CASE STUDIES

Biological assessments, in conjunction with other data (chemical, toxicity, physical, landscape), provide water quality management programs the data and information necessary to document the effectiveness of management actions to protect and restore water quality and to clearly communicate that information to the public. Biological assessment data, WET test results, and physical and chemical monitoring are used to build the relationship between the stressors being managed and the biological impact of the stressors. By relating biological condition to the level and type of stress, results of individual program actions can be related to a common measure of actual environmental improvements—the condition of the aquatic biota (Figure 3-1). The ultimate goal is a water quality management program that integrates biological, physical, and chemical data to create a more complete picture of resource conditions that supports effective implementation of the NPDES and TMDL programs.

By quantifying the stressor-response relationships, it is possible to explain to stakeholders the effects on aquatic life. For example, biological assessment data can be used to document the effects on aquatic life from an undetected toxic effluent from a point source, increasing impervious surfaces in a watershed, the loss of wetlands, or the effects of channelization. Once management actions are implemented, biological assessment data can measure the biological benefits of addressing those effects. That information helps the public understand what is being protected or what could be restored and whether state or tribal water quality standards (WQS) (i.e., aquatic life protection) are being met. Typically, with improved understanding of what is at stake, the public is more informed, motivated, and engaged in working with the state/tribal or local agencies in setting goals for protection or restoration and designing solutions that work.

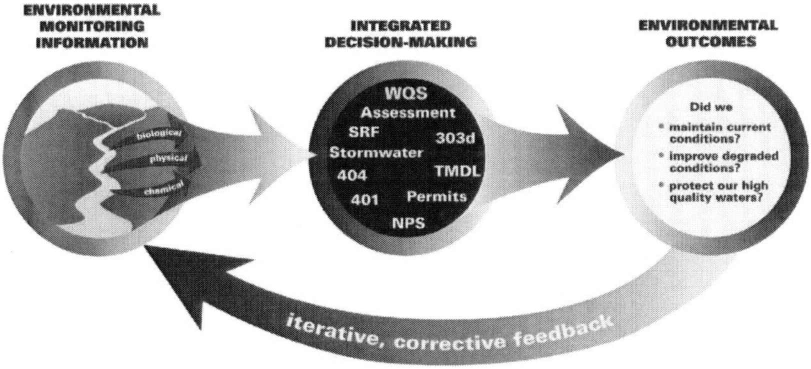

Figure 3-1. Biological data and assessments support integrated decision making.

Over the past four decades, state and tribal water quality programs have used technical tools and information on biological condition to support management decisions. Development of practical methods and technical approaches for biological assessment programs includes field testing by state and tribal programs. These technical advancements build upon existing approaches and can be used by states and tribes to strengthen their biological assessment and biological criteria programs. This Section presents 17 examples of how states and tribes have incorporated such information and tools into their programs or are exploring additional biological condition applications.

The case studies are listed below.

Case Studies

3.1. Protecting Water Quality Improvements and High Quality Conditions in Maine

3.2. Arizona's Development of Biological Criteria

3.3. Protection of Antidegradation Tier II Waters in Maryland

3.4. Using Complementary Methods to Describe and Assess Biological Condition of Streams in Pennsylvania

3.5. Use of Biological Assessments to Support Use Attainability Analysis in Ohio

3.6. Screening Tool to Assess Both the Health of Oregon Streams and Stressor Impacts
3.7. North Fork Maquoketa River TMDL in Iowa
3.8. Addressing Stormwater Flow in Connecticut's Eagleville Brook TMDL for Biological Impairment
3.9. Vermont's Use of Biological Assessments to List Impaired Waters and to Support NPDES
Permit Modification and Wastewater Treatment Facility Upgrades
3.10. Restoration of Red Rock Creek by the Grand Portage Band of Lake Superior Chippewa
3.11. Using Biological Assessment Data to Show Impact of NPS Controls in Michigan
3.12. Using Biological Assessment as Evidence of Damage and Recovery Following a Pesticide Spill in Maryland and the District of Columbia
3.13. Support for Dredge and Fill Permitting in Ohio
3.14. Virginia INSTAR Model for Watershed Protection
3.15. Examination of Climate Change Trends in Utah
3.16. Applications of Biological Assessment at Multiple Scales in Coral Reef, Estuarine, and Coastal Programs
3.17. Partnerships in the Protection of Oregon's Coho Salmon

3.1. Protecting Water Quality Improvements and High Quality Conditions in Maine

Abstract

Maine has used biological, habitat, and other ecological information to designate aquatic life uses that reflect the highest achievable conditions of its waterbodies and has used antidegradation policy to maintain and protect high existing conditions. Maine uses a Biological Condition Gradient to designate levels of protection for its waterbodies (e.g., designated aquatic life uses) and to assign numeric biological criteria to protect those uses.

Maine describes the system as a tiered use classification. For Maine, tiered aquatic life uses highlight the relationship between biology, water quality, and watershed condition in determining the need for waterbody protection to maintain existing high quality conditions or the potential for water quality improvement to attain water quality standards.

Maine's integrated, data-driven approach has resulted in documented improvement in water quality throughout the state, including upgrades of designated uses of more than 1,300 stream miles, from Class C to Class B, and from Class B to Class A or AA waters (Outstanding National Resource Waters).

In 1983 the Maine Department of Environmental Protection (ME DEP) initiated a statewide biological monitoring and assessment program and revised water quality standards (WQS) by 1986 to recognize high levels of water quality condition. Maine established four classes for freshwater rivers and streams (see Table 3-1). All four classes meet or exceed the Clean Water Act (CWA) section 101(a)(2) goal for aquatic life protection. Every waterbody is assigned to one of four tiers by considering its existing biological condition, its highest achievable condition on the basis of biological potential, aquatic habitat, watershed condition, levels of dissolved oxygen, and numbers of bacteria (Table 3-1). Agency biologists developed a linear discriminant model to measure the biological attainment of each class, establish numeric biological criteria, and assign corresponding antidegradation tiers for purposes of statewide planning (see Table 3-1, column 6). Part of Maine's antidegradation policy requires that where any actual measured water quality criterion exceeds that of a higher class, that quality must be maintained and protected [Maine Revised Statutes Title 38, §464.4(F)]. In effect, by having multiple levels of aquatic life use standards in law, Maine has established a means of improving water quality in incremental steps, and of using antidegradation reviews and reclassification upgrades to maintain and protect water quality and aquatic life conditions that exceed existing or designated aquatic life uses.

The following case study offers an example of how Maine has used tiered use classifications and antidegradation policy cooperatively in its water quality management program. In conjunction with habitat and other chemical and physical parameters, Maine assigns waters to designated use classes (AA, A, B, or C; Table 3-1) on the basis of the *potential* for water quality improvement. In the 1980s, monitoring on the Piscataquis River near the towns of Guilford and Sangerville found aquatic life conditions insufficient to meet even the minimum Class C conditions at which the river was classified. The segment of the river in the Guilford-Sangerville area had a history of poor water quality, including recurrent fish kills from poorly treated industrial and municipal wastes. However, the state determined that this segment of the river could attain at least Class C. The state determined that sewage treatment plant and industrial discharges were the only significant source of stressors to the river, with very good quality upstream conditions and good salmonid production elsewhere. Additionally, the river's habitat structure and hydrologic regime were very good.

Table 3-1. Criteria for Maine river and stream classifications and relationship to antidegradation policy

Class	Dissolved oxygen criteria	Bacteria criteria	Habitat narrative criteria	Aquatic life narrative criteria*** and management limitations/restrictions	Corresponding federal antidegradation policy tiers
AA	As naturally occurs	As naturally occurs	Free-flowing and natural	As naturally occurs**; no direct discharge of pollutants; no dams or other flow obstructions.	3 (Outstanding National Resource Water [ONRW])
A	7 ppm; 75% saturation	As naturally occurs	Natural**	Discharges permitted only if the discharged effluent is of equal to or better quality than the existing quality of the receiving water; before issuing a discharge permit the Department shall require the applicant to objectively demonstrate to the department's satisfaction that the discharge is necessary and that there are no reasonable alternatives available. Discharges into waters of this class licensed before 1/1/1986 are allowed to continue only until practical alternatives exist.	2 1/2
B	7 ppm; 75% saturation	64/100 mg (g.m.) or 236/100 ml (inst.)*	Unimpaired**	Discharges shall not cause adverse impact to aquatic life** in that the receiving waters shall be of sufficient quality to support all aquatic species indigenous** to the receiving water without detrimental changes to the resident biological community.**	2 to 2 1/2

Class	Dissolved oxygen criteria	Bacteria criteria	Habitat narrative criteria	Aquatic life narrative criteria*** and management limitations/restrictions	Corresponding federal antidegradation policy tiers
C	5 ppm; 60% saturation; and 6.5 ppm (monthly avg.) when temperature is </= 24 °C	125/100 mg (g.m.) or 236/100 (inst.)*	Habitat for fish and other aquatic life	Discharges may cause some changes to aquatic life**, provided that the receiving waters shall be of sufficient quality to support all species of fish indigenous** to the receiving waters and maintain the structure** and function** of the resident biological community. **	1 to 2

Source: Maine DEP (modified). http://www.maine.gov/dep/blwq/docmonitoring/classification/reclass/appa.htm.

Notes:

* g.m. = geometric mean; inst. = instantaneous level.

** Terms are defined by statute (Maine Revised Statutes Title 38, §466).

*** Numeric biological criteria in Maine regulation Section 579, Classification Attainment Evaluation Using Biological Criteria for Rivers and Streams.

Four years after issuance of new National Pollutant Discharge Elimination System (NPDES) permits requiring better industrial pretreatment and improved wastewater treatment at the Guilford-Sangerville treatment facility, follow-up monitoring found water quality improvements that exceeded Class C and attained Class B aquatic life conditions. The achievement of higher water quality conditions was preserved through a classification upgrade process (supported by the industry and the two towns). The river was upgraded to Class B and now attains those higher aquatic life use goals. The redesignation process requires the state legislature to enact a statutory change of a waterbody's classification and can take considerable time to complete. However, during the reclassification process the improved water quality conditions existing in the Piscataquis River were protected through implementation of the state's Tier II antidegradation policy. The value secured by maintaining the higher quality condition was demonstrated in 2009 when the Piscataquis River was designated as critical habitat for the restoration of the endangered Atlantic salmon.

The management actions based on documented improvements in the biological condition in this example demonstrate the complementary application of the state's tiered aquatic life use classification and the Tier 2 and 21/2 antidegradation policy. Using that approach, water quality upgrades from Class C to B and from B to A or AA have been repeated in many parts of the state, and subsequently maintained and protected. Overall, Maine has redesignated more than 1,300 miles of streams to a higher class on the basis of biological information (e.g., biological improvements due to point source controls, nonpoint source practices, dam operational modifications or removal) and societal values (e.g., water quality and habitat protection for wild trout populations; critical species protection, especially Atlantic salmon habitat and tribal petitions).

3.2. Arizona's Development of Biological Criteria

Abstract

Arizona has adopted in its water quality standards both narrative and numeric biological criteria to help protect aquatic life uses in wadeable, perennial streams designated for either coldwater aquatic and wildlife or warmwater aquatic and wildlife. The biological criteria allow the state to define expected conditions relative to reference streams. The state implements a two-step verification process to confirm attainment of the biological criteria for waters that score just below the attainment threshold. Arizona Department of Environmental Quality uses the biological assessment results in its 305(b) reports on the condition of its aquatic resources.

Development of Numeric Biological Criteria

Arizona began a biological assessment program in 1992, following EPA's Rapid Biological Assessment Protocols for wadeable streams and rivers (Plafkin et al. 1989). Standard operating procedures for macroinvertebrate monitoring in perennial, wadeable streams and for laboratory processing and taxonomic identification were established and have been periodically reviewed and updated (ADEQ 2010). A statewide reference monitoring network was established to develop an index of biological integrity (IBI) as the macroinvertebrate assessment method.

A classification analysis was first performed on the statewide macroinvertebrate data set to identify regions of statistically different macroinvertebrate communities across the state (Spindler 2001). Elevation-based regions were the result of the classification analysis, consisting of two broad macroinvertebrate regions and community types:

- A warmwater community below 5,000 feet elevation
- A coldwater community above 5,000 feet elevation

All wadeable, non-effluent-dependent perennial streams in the regions, with some documented exceptions, are predicted to have the same general macroinvertebrate community type. IBIs were then developed for both a warmwater and coldwater community using the statewide reference site data (ADEQ 2007).

In the initial stages of development, Arizona's numeric biological criteria were based on the idea that the structure and function of aquatic benthic macroinvertebrate communities provide information on the overall quality of their surface waters and on attainment of the state's designated aquatic life uses. Measuring the composition and structure of the biological communities in minimally disturbed surface waters provides reasonable approximation of biological integrity and, thus, the basis for establishing the reference condition (Stoddard et al. 2006). The reference condition provides the benchmark for evaluating the biological condition of surface waters that could have been subjected to relatively greater amounts of disturbance.

However, on the basis of the state's scrutiny of the reference site database and further investigation of surrounding land use, the state concluded that its reference sites represent *best available*, or *least disturbed*, conditions for each watershed. There was uncertainty as to whether some of the reference sites at the lower range of the reference distribution were truly minimally disturbed conditions. For example, while reference sites were in a wilderness area for streams considered to be in pristine condition, much of the watershed upstream was extensively grazed, and the index scores for the reference sites were lower than the mean. In addition, there was variability because of sites later found to be intermittent in flow, and samples were affected by extreme flooding in the reference data set. Because of that uncertainty in reference quality in the low end of the reference database, Arizona selected the 25th percentile of the reference site distribution to be protective of the aquatic life use.

Minimally Disturbed Condition: The physical, chemical, and biological conditions of a waterbody with very limited human disturbance compared with natural, undisturbed conditions. There might be some changes to the composition of the resident aquatic biota, but native species are present.

Least Disturbed Condition: The best existing physical, chemical, and biological condition of a waterbody affected by human disturbance.

These waters have the least amount of human disturbance in comparison to others within the waterbody class, region, or basin. Least disturbed condition is a relative term, and the actual condition may depart significantly from natural, undisturbed conditions or minimally disturbed conditions. Least disturbed condition might change significantly over time as human disturbances change.

Arizona established a two-stage process for determining nonattainment of the numeric biological criteria. On the basis of statistical analysis of reference, stressed, and test data sets, an attainment threshold of 25 percent was selected. The nonattainment biological criteria threshold was set at the 10th percentile of reference, the level at which a majority of stressed samples occurs in the Arizona Department of Environmental Quality (ADEQ) database. An inconclusive zone falls between the 10th and 25th percentiles of reference. The zone of uncertainty encompasses variability in IBI scores near the 25th percentile. To verify the biological integrity of the *inconclusive* samples, verification sampling is required before making an attainment decision. Verification monitoring must be conducted during the next immediate spring or fall index period. (A fall-based IBI scoring system is being developed.) If the waterbody in question scores at or less than the 25th percentile of reference, it will then be judged as not attaining. Such a verification approach provides an opportunity to confirm the status of waters that score just below the attainment threshold of the biological criteria.

Adoption of Numeric Biological Criteria

On January 31, 2009, Arizona adopted biological criteria, as part of the revised Arizona surface water quality standards (WQS), applicable to wadeable, perennial streams with either a coldwater or warmwater designated aquatic life use. The biological criteria consist of two parts: a narrative statement (Arizona R18-11-108) and numeric criteria (ARS R18-11-108.01). The narrative is presented as follows:

> A wadeable, perennial stream shall support and maintain a community of organisms having a taxa richness, species composition, tolerance, and functional organization comparable to that of a stream with reference conditions in Arizona.

The numeric criteria are laid out in text and numeric form (Table 3-2) in the state's biological criteria rule in the WQS as follows:

The biological standard in R18-11-108(E) is met when a biological assessment result, as measured by the Arizona IBI [index of biological integrity], for cold or warm water is: 1) Greater than or equal to the 25th percentile of reference condition, or 2) Greater than the 10th percentile of reference condition and less than the 25th percentile of reference condition and a verification biological assessment result is greater than or equal to the 25th percentile of reference condition.

Table 3-2. Arizona numeric biological criteria IBI scores

Biological assessment result	IBI scores	
	coldwater	warmwater
Greater than or equal to the 25th percentile of reference condition	> 52	> 50
Greater than the 10th and less than the 25th percentile of reference condition	46–51	40–49

Source: Arizona R18-11-108.01.

ADEQ uses the biological assessment results in its 305(b) reports on the condition of its aquatic resources. More information about the biological criteria, sampling methods, establishing reference condition, and the method for determining nonattainment of the biological criteria is provided in *Biocriteria Implementation Procedures* (ADEQ 2008) and in *Technical Support Documentation for the Narrative Biocriteria Standard* (ADEQ 2007).

3.3. Protection of Antidegradation Tier II Waters in Maryland

Abstract

Maryland is identifying high-quality waters for antidegradation purposes on a waterbodyby-waterbody basis. Maryland has designated Tier II waters on the basis of two indices of biotic integrity—fish and benthic invertebrates—and provides additional protection so that those waters are not degraded. New or increased point source dischargers and local sewer planning activities that have the potential to affect Tier II waters are required to examine alternatives to eliminate or reduce discharges or impacts. The state has developed requirements that must be met for projects that do not implement a no-discharge alternative.

To help local planners to determine whether a planned activity has the potential to affect a Tier II water, the state has developed geographic information system shapefiles that identify such waters. Those files are provided to local jurisdictions to improve their knowledge of where Tier II waters occur. Biological assessments, in conjunction with chemical and physical assessments, are then conducted to determine the status of those waters and detect trends in condition.

In its state water quality standards (WQS), Maryland adopted an antidegradation policy for protecting all waters for existing and designated uses. High-quality (Tier II) waters receive additional attention and regulatory protections. Identification of Tier II waters, in this case streams, is based on a waterbody-bywaterbody approach using biological survey data, from which two indices of biotic integrity (IBIs) are developed—one for benthic invertebrates and one for fish. Those with both scores above 4 are designated Tier II waters. The state has identified more than 230 high-quality water segments. To protect downstream high-quality waters, a watershed approach to protection is applied. Tier II waters must be protected so that water quality does not degrade to minimum standards, and that requirement has implications for potential discharges and local planning activities.

Application of Tier II Protection

The Maryland Department of the Environment (MDE) requires that applicants for amendments to county plans (i.e., water and sewer plans) or permits for new or expanding point source discharges evaluate alternatives to eliminate or reduce discharges or impacts [COMAR 26.08.02.04-1(B)]. Applicants for permits must consider whether the receiving waterbody is Tier II (or whether a Tier II determination is pending); MDE reviews proposed amendments to county plans discharging to Tier II waters. In both cases, discharges to Tier II waters require a Tier II review [2.26.08.02.04-1(F)].

MDE has developed a cooperative approach to protecting Tier II waters. Monitoring and WQS programs work with the National Pollutant Discharge Elimination System (NPDES) permitting program to help screen for potential effects from new or expanded discharges and to develop permit conditions to minimize those effects and maintain existing high-quality waters. Outreach materials are available to educate county planners about Tier II waters, and geographic information system (GIS) shapefiles that planners can use to help locate Tier II waters within their jurisdictions have been developed.[15] That information provides Maryland county planners a way to determine early on whether their projects could affect Tier II waters.

A list of recommendations for land-disturbing projects that are not able to implement a no-discharge alternative provides the following initial guidance:

1. Implementation of environmental site design (also known as low-impact development)—Design elements and practices must be approved for Tier II waters with opportunity provided for exploration of appropriate alternatives and justification for structural elements in the proposed designs.
2. Expanded riparian buffers—Buffers must be at a minimum of 100 feet; wider buffers may be required depending on slope and soil type.
3. Biological, chemical, and flow monitoring in the Tier II watershed—Applicants may be required to conduct biological assessments in conjunction with chemical, physical, and flow assessments to help determine the remaining assimilative capacity and cumulative impacts of current and future development. Depending on project specifics, additional monitoring may be required, such as the completion of a hydrogeologic study for a major mining project or additional pH monitoring because of impacts associated with instream grout applications seen in many common transportation projects.
4. Additional practices—Depending on the potential for project specific effects on water quality, applicants may be required to implement other practices, such as enhanced sediment and erosion control practices or implementation of more environmentally protective alternatives.

If those general requirements cannot be implemented, applicants must submit a detailed hydrologic study and alternatives analysis to demonstrate that the assimilative capacity of a waterbody will be maintained. The assimilative capacity of a waterbody is typically site-specific and determined through studies of the waterbody. In terms of WQS, assimilative capacity is a measure of the capacity of a receiving water to assimilate additional pollutant(s) but still meet the applicable water quality criteria and designated uses.

3.4. Using Complementary Methods to Describe and Assess Biological Condition of Streams in Pennsylvania

Abstract

The Pennsylvania Department of Environmental Protection (PA DEP) has developed a new benthic macroinvertebrate index of biotic integrity (IBI) to assess the health of wadeable, freestone (e.g., high gradient, soft water) streams. Additionally, PA DEP calibrated a benthic macroinvertebrate Biological Condition Gradient (BCG) and is exploring using the BCG to more precisely describe biological characteristics in Pennsylvania streams. Potentially, the BCG can be used in conjunction with the IBI to identify aquatic life impairments and to describe the biological characteristics of waters assigned special protection. PA DEP is also exploring using a discriminant analysis model with additional taxonomic, habitat, and landscape parameters to describe exceptional value waters.

Describing Waters along a Gradient of Condition

Pennsylvania Department of Environmental Protection (PA DEP) has developed a new benthic macroinvertebrate index of biotic integrity (IBI) for the wadeable, freestone (high-gradient, soft-water) streams in Pennsylvania using the reference condition approach (PA DEP 2009). PA DEP has alternative assessment methods in place for other stream types (i.e., low-gradient pool-gliders, karst [limestone]-dominated). The IBI provides an integrated measure of the overall condition of a benthic macroinvertebrate community by combining multiple metrics into a single index value. PA DEP uses the IBI to assess attainment of aquatic life uses.

Additionally, PA DEP is exploring use of a Biological Condition Gradient (BCG) to describe the biological characteristics of freestone streams along a gradient of condition. PA DEP conducted a series of three expert workshops in 2006, 2007, and 2008 to calibrate a BCG along a gradient from minimally to heavily stressed conditions (PA DEP 2009). The BCG is a narrative model based on measurable attributes, or characteristics, of aquatic biological communities expected in natural conditions (e.g., presence of native taxa, some pollution tolerant taxa present but typically not dominant, absence of invasive species). Additionally, the BCG model includes attributes that describe interactions among biotic communities (e.g., food web dynamics), the spatial and temporal extent of stress, and the presence of naturally occurring habitats and landscape condition (for more information, see Tool # 2, *The Biological Condition Gradient*). To date, states and tribes that have applied the BCG have used the BCG attributes that describe the taxonomic composition of the resident aquatic biota and, where available, information on fish condition, for example lesions and abnormalities (BCG attributes I–VII) (see Table 2-2). Some states are exploring the application of additional attributes on food web dynamics, extent of stress, and landscape condition (BCG attributes VIII–X). These efforts are providing valuable information that will aid the U.S. Environmental Protection Agency (EPA) in further refining the BCG.

> A **metric** is a measurable aspect of a biological community that responds in a consistent, predictable manner to increasing anthropogenic stress.
>
> Examples of metrics include **taxa richness**, which is a measure of the number of different kinds of organisms (taxa) in a sample collection, and **% dominance**, which is a measure of which species compose the majority of organisms present in a sample collection.
>
> To gain a more comprehensive view of an aquatic community, multiple types of metrics are combined into a **biological, or biotic, index**. The typical biological index may include information from 7 to 12 different metrics. The metric values are typically scored on a unitless scale of 0 to 100 and averaged to obtain a single value.

To develop the BCG for its streams, biologists from PA DEP, in conjunction with external taxonomic experts and scientists, e.g., the Delaware River Basin Commission, Western Pennsylvania Conservancy, and EPA, used the BCG attributes that characterize specific changes in community taxonomic composition (PA DEP 2009). For example, in the highest tiers of the BCG, locally endemic, native, and sensitive taxa are well represented (attributes I and II) and the relative abundances of pollution-tolerant organisms (attribute V) are typically lower. With increasing stress, more pollution-tolerant species may be found with concurrent loss of pollution-sensitive species (attribute VI). At the beginning of the expert workshops, the biologists first assigned or adjusted BCG attributes to each macroinvertebrate taxon (e.g., pollutant-sensitive or tolerant) and then reviewed taxa lists from samples representing minimally disturbed to severely disturbed site conditions (Figure 3-2). The evaluated samples included sites judged as either reference quality (e.g., at or close to minimally disturbed conditions) or heavily stressed based on specific selection criteria (PA DEP 2009). To further test the robustness of the BCG process, additional sites that were not part of the reference or heavily stressed sample groups were evaluated. Those sites represented a range of site conditions, including moderately to heavily stressed site conditions (non-reference and moderately stressed; see Figure 3-2). Using the BCG tier descriptions of predicted changes in the attributes as a guide, they assigned each site to one of the six BCG tiers.

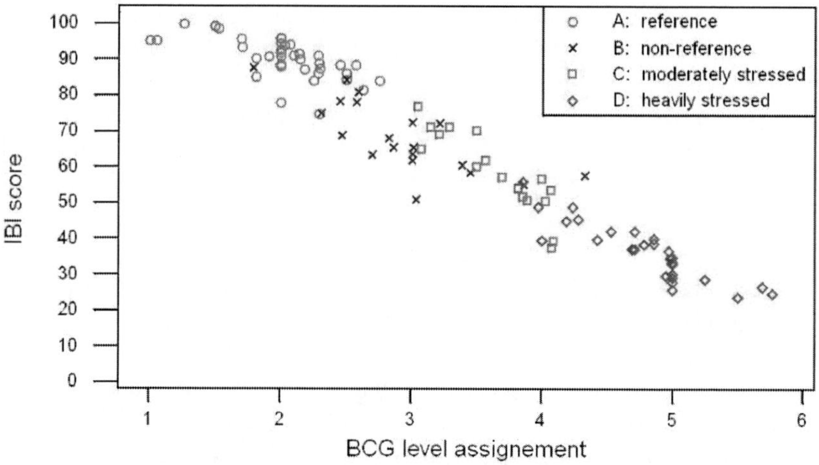

Figure 3-2. Comparison of calibrated BCG tier assignments (mean value) and IBI scores for freestone streams representing range of conditions from minimal to severely stressed.

For all the evaluated samples, PA DEP biologists analyzed the relationship between a sample's BCG tier assignment with its corresponding IBI score (PA DEP 2009). A strong correlation existed between the calibrated BCG tier assignments and the IBI scores (Figure 3-2). Based on these results, PA DEP is evaluating using the BCG to describe the biological characteristics of streams along a gradient of condition; for example, the reference sites clustered at IBI scores near 80 and above. Based on taxonomic information and without knowledge of the IBI scores, the experts assigned these sites to BCG tiers 1.5 to 2.5. BCG tier 2 represents close to natural conditions (e.g., minimal changes in structure and function relative to natural, or pristine, conditions; supports reproducing populations of native species of fish and benthic macroinvertebrates). This information can meaningfully convey to the public the biological characteristics of waters in the context of the Clean Water Act and the goal to protect aquatic life. Using both the IBI and BCG, PA DEP might be able to develop a cost-effective, publicly transparent approach to routinely monitor and assess the condition of its freestone streams and to help identify potential high-quality (HQ) or exceptional value (EV) streams.

Describing Exceptional Value Waters

Pennsylvania's regulations define waters of EV that are of unique ecological or geological significance. EV streams are given the highest level of protection and constitute a valuable subset of Pennsylvania's aquatic resources. To support protection of these waters, PA DEP is considering the use of a discriminant analysis model to evaluate the relationship between condition of the watershed, a stream, and its aquatic biota (e.g., the connection of riparian areas with a stream and the floodplain or the spatial extent of stressors and their sources in the watershed). PA DEP is evaluating the use of a discriminant model that incorporates measures of land use and physical habitat along with IBI scores and indicator taxa richness to make distinctions between EV and HQ waters. The abiotic measures PA DEP is using address habitat fragmentation and spatial and temporal extent of stress and are comparable to the national BCG model attributes IX (extent of stress) and X (ecosystem connectance). The results of this effort could potentially support decisions on where to target resources for sustainable, cost-effective protection of EV waters and healthy watersheds. Through this work, PA DEP is providing EPA valuable feedback on the technical development and potential program application for BCG attributes IX and X.

Potential Application to Support Protection of Waters of Highest Quality

PA DEP is exploring new approaches to help identify streams that are of the highest quality and might require special protection. For example, a stream might be found to meet the expected biological condition of an HQ or EV water based on its IBI score and BCG tier assignment. This information could be used to support further study to determine whether its designation should be as an HQ water or if it meets the additional criteria for designation as an EV water. When biological information is presented in context of a BCG framework, it is easier for the public to understand the status of the aquatic resources, including waters that are in excellent condition and require additional protection.

3.5. Use of Biological Assessments to Support Use Attainability Analysis in Ohio

Abstract

Ohio uses biological assessment information in conjunction with physical habitat assessments to strengthen use attainability analyses (UAAs) in the state. The technical and programmatic underpinnings for Ohio's use attainability determinations is the state's aquatic life use classification approach, which is based on the relationship between biology, habitat, and the potential for water quality improvement. Ohio's biological monitoring and assessment program provides timely, statewide information on the status of waterbodies and the data to support a UAA if needed, including when biological conditions improve and an upgrade of a designated use is warranted. Typically, in situations where the habitat needed to meet aquatic life uses is present, Ohio has taken management actions to address water quality issues and restore impairments.

In 1990 Ohio used biological assessment information to specify levels of biological condition for specific streams and rivers based on ecoregional reference sites. As a result, the state refined definitions of some aquatic life uses, adopted new ones, and assigned biological criteria to key uses to support a tiered approach to water quality management within the Ohio water quality standards (Table 3-3).

When designating aquatic life uses, the quality of habitat is a major factor in a use attainability analysis (UAA) process to determine the potential for restoration and expected biological condition for streams and rivers in Ohio. If sufficient good habitat attributes are not present, such as higher quality substrates and sufficient instream cover, a determination about restorability is made. If habitat is sufficient or could be restored, it is assumed that any observed biological impairments are due to the effects of other stressors (e.g., metals, nutrients) that could be remediated through readily available water quality management options (e.g., permit conditions and/or best management practices [BMPs]) and the biological assemblage restored. The aquatic life use classifications are based on ecological conditions, and in 1990 biological criteria were developed to protect each use. Ohio's biological criteria include two indices based on stream fish assemblages (Index of Biological Integrity [IBI] and Modified Index of Well-Being [MIwb]) and one index based on stream macroinvertebrate assemblages (Invertebrate Community Index [ICI]).

The biological criteria were developed based on regional reference conditions and are stratified by each of the state's five level 3 ecoregions and three site types (headwater, wadeable, and boatable sites).

Table 3-3. Summary of Ohio's beneficial use designations for the protection of aquatic life in streams

Beneficial use designation	Key attributes
Coldwater habitat (CWH)	Native cold water or cool water species; put and take trout stocking.
Exceptional warmwater habitat (EWH)	Unique, unusual, and highly diverse assemblage of fish and invertebrates.
Seasonal salmonid habitat (SSH)	Supports lake run steelhead trout fisheries.
Warmwater habitat (WWH)	Typical assemblages of fish and invertebrates, similar to least impacted reference conditions.
Limited warmwater habitat (LWH)	Temporary designations based on 1978 WQS. Predate Ohio tiered aquatic life use classification and were not subjected to UAA; being phased out as UAA are conducted for each LWH waterbody or segment. Most of the LWH waterbodies or segments have been redesignated as WWH or higher with the exception of some mine-drainage-affected segments that were designated LRW.
Modified warmwater habitat (MWH)	More tolerant assemblages of fish and macroinvertebrates are present relative to a WWH assemblage, but otherwise generally similar species to WWH present; irretrievable modifications of habitat preclude complete recovery to least impacted reference condition.
Limited resource water (LRW)	Fish and macroinvertebrates severely limited by physical habitat or other irretrievable condition; minimum protection afforded by the CWA.

Source: Ohio EPA, April 2004. http://www.epa.ohio.gov/portals/35/wqs/designation summary.pdf.

Using these aquatic life use classifications, Ohio has been able to determine attainable levels of condition for streams and rivers. For example, in the mid-1980s biological surveys of Hurford Run, a small stream located in an urban/industrial area of Canton, Ohio, showed that the stream was severely impaired by toxic chemical pollutants and that some sites had no fish at all. Hurford Run is channelized for nearly its entire length. Because of the severity of the biological impairment, a UAA was conducted to determine if the warmwater habitat (WWH) aquatic life use was attainable and, if not, to determine the most appropriate designated use for the stream. Based on biological and habitat assessments, the most appropriate aquatic life uses for the different segments of Hurford Run could be determined. For example, very poor habitat quality from historical channelization in the *upper reach of Hurford Run* and the associated hydrological modifications (e.g., ephemeral flows) resulted in a limited warmwater habitat (LWH) designation for this upper reach.

The *middle reach of Hurford Run* has been subject to extensive, maintained channel modifications that also resulted in degraded habitat features, though water is always present. Channel maintenance practices resulting in poor-quality substrates, poorly developed pools and riffles, and a lack of instream cover preclude biological recovery to assemblages consistent with the WWH use, which indicated that the middle reach should be designated a modified warmwater habitat (MWH), reflecting the attainable biological potential for a channel-modified stream determined by scientific studies. The *lower reach of Hurford Run* was previously relocated and channelized, but over time the reach has naturally recovered sufficient good-quality habitat attributes, such as coarse substrates and better developed riffle and pool features associated with the WWH use for this ecoregion. Biological assessments confirmed the presence of aquatic assemblages typical of WWH. Based on this information, this segment was designated as WWH. The designated aquatic life uses reflect the current best possible condition in each segment of Hurford Run and provide a basis for management actions to ensure that the associated criteria are met and the use is protected. Numeric biological criteria have been established for key designated aquatic life uses, and a segment is listed on the 303(d) list if it is in nonattainment of the biological criteria. Additionally, the different segments are routinely monitored by the state and the condition reevaluated on a regular basis. If there is any information indicating that a higher use is being attained or could be attained, that water is considered for redesignation to the higher use.

Ohio has also used biological assessment data to refine its water quality criteria in some cases. For instance, when Ohio's aquatic life use classifications were established in 1978, Ohio established dissolved oxygen criteria to protect each designated use. Initially, a dissolved oxygen criterion of 6 mg/L as a minimum was established for exceptional warmwater habitat (EWH) waters to protect highly sensitive species supported by this use. However, analyses of ambient biological and chemical data suggested that the

6 mg/L minimum criterion was over-protective for EWH waters. Data showed a relationship between stressors and biological measures, with dissolved oxygen concentrations less than 5.0 mg/L being associated with IBI scores not in attainment of EWH biological criteria. And, in general, data showed that with dissolved oxygen greater than 5.0 mg/L, IBI scores are much more likely to attain EWH. These results were used to justify refining the EWH criteria to the current 6 mg/L average, 5 mg/L minimum (Ohio EPA 1996). The criterion revision also supported the redesignation of some rivers and streams from WWH to EWH.

3.6. Screening Tool to Assess Both the Health of Oregon Streams and Stressor Impacts

Abstract

The Oregon Department of Environmental Quality conducted a study in the John Day River Basin to both evaluate the biological health of streams using biological sampling for macroinvertebrates and to identify the causes of stream impairment using biological monitoring information. The state used the PREDATOR model to evaluate waterbody conditions in perennial streams. Stressor identification models were used to measure the effects of stress from two sources of nonpoint source pollution (excessive temperature and fine sediment). A comparison of modeling results to sampling data showed that both modeling and direct measurements are useful in identifying streams not meeting benchmarks and identifying cause of impairment. Oregon will continue to use the model results to evaluate the ability to identify causes of biological impairment on the basis of macroinvertebrate data and will use that information to improve water quality.

The John Day River Basin in northeastern Oregon is one of the state's most important scenic waterways. It drains nearly 8,100 square miles of land and is one of the nation's longest free-flowing river systems (BLM 2010). Oregon Department of Environmental Quality (ODEQ) evaluated the biological health of streams in the John Day River Basin using biological sampling for macroinvertebrates. The study also identified the causes of stream impairment with the aid of biological monitoring information. The focus of the studies conducted by ODEQ was to model the biological condition and explore the relative importance of the two most common nonpoint source (NPS) stressors—elevated temperature and excess fine sediments—using macroinvertebrate data.

Biological Condition Model (PREDATOR)

ODEQ sampled benthic macroinvertebrates in 76 perennial, wadeable streams in the John Day River Basin. The biological condition of the streams was modeled using ODEQ's PREDictive Assessment Tool for ORegon (PREDATOR) (Hubler 2008). The model predicts the kinds of macroinvertebrates expected to occur at reference sites with similar environmental conditions (precipitation, air temperature, elevation, and ecoregion). For example, high-elevation sites that experience higher precipitation levels and cooler air temperatures in eastern Oregon would be expected to support macroinvertebrates similar to those found at reference sites that are both geographically and environmentally similar.

The PREDATOR model uses 176 reference sites across five Level III ecoregions in Oregon (Omernik 1987). The model output is the ratio of the macroinvertebrates observed at a test site (O) to the expected macroinvertebrates (E), or O/E. Values less than 1.0 represent a loss of reference macroinvertebrates at the test site relative to natural conditions. ODEQ classifies sites into one of three biological condition classes: *least disturbed*, *moderately disturbed*, and *most disturbed*. Oregon's least disturbed class supports native populations of aquatic macroinvertebrates and natural habitat.

The results of the study indicated that almost half of the sites were in least disturbed conditions, or equivalent to reference (O/E values close to 1.0). Just over one-quarter (28 percent) were in most disturbed conditions with O/E values down to 0.47, indicating loss of over half of the expected, or native, species.

NPS Pollutant Stressor Models

To use macroinvertebrates to measure the effects of stress from NPS pollution (temperature and fine sediments), ODEQ used two *stressor identification* (SI) models (Huff et al. 2006). Temperature stress (TS) and fine sediment stress (FSS) are two new biological indices used to infer seasonal maximum temperature and percent fine sediments based entirely on the macroinvertebrates collected at a site.

Those indices consistently and predictively respond to increased levels of temperature or fine sediments and are used to model macroinvertebrate-specific changes to the stressors (e.g., stressor-response signatures).

Comparisons of Stressor Model Output to Field Measurements

Water quality and physical habitat information was also collected as part of the John Day River Basin study. Direct comparisons of the SI models (assemblage response signatures) to their equivalent physical measurements (water column temperature and fine sediment load) show similar abilities in determining the extent of streams failing to meet benchmarks. However, the SI models showed a stronger relationship to biological condition than did the physical measurements of temperature and fine sediments. Most of the test sites in good condition according to the PREDATOR model coincided with the SI model outcomes also in good condition. The test sites in good biological condition supported specific macroinvertebrates with temperature and fine sediment preferences similar to reference assemblages. Conversely, the majority of sites in poor biological condition (most disturbed) had TS and FSS values above the reference benchmark for the SI model. To further identify the relative importance of temperature and fine sediments to biological condition, ODEQ routinely performs more quantitative analyses. Regression models of the relationship between PREDATOR and SI models can be used to identify the strength and significance of relationships. Additionally, relative risk analysis is used to quantitatively rank the importance of stressors to biological condition.

Conclusion

ODEQ developed two SI models that can be used to identify the relative importance of two common NPS stressors— elevated temperatures and fine sediments—to biological condition. ODEQ's primary objective with the analysis was to explore the ability of macroinvertebrate data to identify causes of biological impairment.

The results from the study show that about one-half of the perennial, wadeable streams in the John Day River Basin are in good condition, one-quarter are in fair condition, and one-quarter are in poor condition. SI models were used to identify primary causes of biological impairment from NPS pollution. Although biological measures and physical measures were comparable in their ability to detect the extent of sites with NPS stressors above levels typically observed at reference sites, the biological measures showed a stronger relationship to biological condition.

The models for biological condition and SI show promise as sensitive and cost-effective screening tools for detecting NPS impairment to streams and targeting best management practices (BMPs) to address the primary stressors, elevated temperature and excessive fine sediment loads. The SI models also provide benchmarks to measure the response of the biological community to BMP implementation. Combining the information from the models can help scientists better understand the risks associated with NPS pollution in Oregon streams and more efficiently target resources to improve water quality.

Complementary Application of Biological Condition and Stressor Identification Models: An Example

Biological Condition: North Fork Deer Creek had a list of 14 expected macroinvertebrate taxa that were frequently observed at reference sites with similar geographical and environmental characteristics. However, only nine of the expected taxa were observed at the sampling site, resulting in a rating of most disturbed condition (O/E = 9/14 = 0.64).

Stressor Identification: The SI models were used to infer temperature and fine sediment conditions using the tolerances and abundances of all macroinvertebrates collected at North Fork Deer Creek. The dominant macroinvertebrates in the creek showed high tolerances to fine sediments, while the same taxa showed preferences for cooler water over warmer water. For example, five taxa were indicators (taxa that exhibit strong preferences) of higher fines at a site, compared to one indicator taxa for low fines. Additionally, five taxa were indicators of cool water conditions in North Fork Deer Creek, compared to one indicator taxa of warmwater conditions. The tolerances of the most abundant macroinvertebrates observed in North Fork Deer Creek indicate that excess fine sediments are the most likely cause of the poor (most disturbed) biological condition.

3.7. North Fork Maquoketa River TMDL in Iowa

Abstract

In 1998 the Iowa Department of Natural Resources (IDNR) determined that a 19.5-mile segment of the North Fork Maquoketa River (NFMR) was not meeting its aquatic life use due to a biological impairment of "unknown cause."

This determination was based on biological assessments of benthic macroinvertebrate and fish populations. All collected data were used by IDNR in the development of a stressor identification (SI) process that showed that the primary pollutants in the NFMR were sediment, nutrients (specifically phosphorus), and ammonia. In 2007 IDNR completed a Total Maximum Daily Load report for the NMFR that used results of the SI process and calls for steep reductions in sediment reaching the river and in nutrients and agricultural manure releases. IDNR also identified a variety of best management practices to improve water quality and is encouraging local residents and businesses to take action to restore their watershed.

Water Quality Impairment of the North Fork Maquoketa River

The North Fork Maquoketa River (NFMR) is designated by Iowa for aquatic life protection as a class B (WW-2)[16] water. In 1998 the NFMR was determined not to be meeting its aquatic life uses based on biological assessments of the benthic macroinvertebrate population that showed low total abundance and species diversity and several reported fish kill events of unknown source. Iowa subsequently placed the 19.5-mile segment that extends from the headwaters near Luxemburg to Dyersville on its 1998 Clean Water Act (CWA) section 303(d) list of impaired waters. The segment was listed for a biological impairment of "unknown causes" (IDNR no date).

Monitoring and Stressor Identification

Iowa Department of Natural Resources (IDNR) conducted additional biological monitoring of the NFMR between 1999 and 2005. Data collection included the number of benthic macroinvertebrates (by lowest practical taxon), number of fish (by species), and instream and riparian habitat assessments. IDNR used these data to calculate a Benthic Macroinvertebrate Index of Biotic Integrity (BMIBI) and a Fish Index of Biotic Integrity (FIBI) that quantify several aquatic community characteristics such as relative abundance of sensitive and tolerant species, and the proportion of organisms belonging to various feeding, spawning, or habitat classifications. BMIBI and FIBI scores for the NFMR watershed are provided in Table 3-6. For the sites sampled, the BMIBI and FIBI ranged from poor to fair (Table 3-4). None of the BMIBI or FIBI scores attained the reference biological criteria (Table 3-5). Qualitative scoring guidelines for the BMIBI and FIBI are summarized in Table 3-4, while reference values are included in Table 3-5.

Table 3-4. Qualitative scoring guidelines for the BMIBI and FIBI

Biological Condition Rating (BCR)	BMIBI	FIBI
Poor	0-30	0-25
Fair	31-55	26-50
Good	56-75	51-70
Excellent	76-100	71-100

Source: http://www.iowadnr.gov/Environment/WaterQuality/WatershedImprovement /WatershedResearchData/ Water ImprovementPlans/PublicMeetingsPlans.aspx. (Note that the NFMR TMDL .pdf document is available under the heading "Final Water Quality Improvement Plans.")

Table 3-5. Reference criteria for assessing biological integrity

Ecoregion[a]	BMIBI	FIBI
52B Ref. (Paleozoic Plateau)	61	59
47C Ref. (Iowan Surface)	59	71 (riffle), 43 (non-riffle)

Source: http://www.iowadnr.gov/Environment/WaterQuality/Watershed Improvement /WatershedResearchData/ Water ImprovementPlans/PublicMeetingsPlans.aspx. (Note that the NFMR TMDL .pdf document is available under the heading "Final Water Quality Improvement Plans.")

[a] The watershed contributing to flow in the NFMR upstream from Dyersville, Iowa, is a transitional area that is divided between two ecological regions of Iowa. Roughly two-thirds of the lower portion of the watershed is located in the Iowan Surface of the Western Corn Belt Plains, while the upper third is located in the Paleozoic Plateau (Driftless Area) ecoregion.

In addition to biological monitoring, IDNR also collected monthly water quality samples in 2001 and 2005 (March through November) for several chemical and physical parameters, such as flow, dissolved oxygen, temperature, pH, nitrate + nitrite, and total phosphorus. The data showed water quality impacts relative to levels measured at least disturbed ecoregion reference stream sites—especially elevated concentrations of ammonia, nitrate-nitrogen, total phosphorus, and total suspended solids. Occasional violations of dissolved oxygen criteria were found, and large diurnal fluctuations in dissolved oxygen concentrations in the stream were indicative of elevated primary production levels. All collected biological, chemical, and physical data were used in the stressor identification (SI) process (IDNR 2006).

Table 3-6. BMIBI and FIBI results for the NMFR Watershed. (BCR rating in parenthesis)

Site	Year	BMIBI	FIBI
REMAP 147	2005	42 (Fair)	34(Fair)
TMDL 28	2001	47 (Fair)	29 (Fair)
TMDL 28	2005	26 (Poor)	37 (Fair)
New Wine Park	1999	N/A[a]	32
TMDL 29	2001	47 (Fair)	26 (Fair)
TMDL 30	2001	51 (Fair)	33 (Fair)
TMDL 30	2005	48 (Fair)	7 (Poor)
H2	1999	53 (Fair)	37 (Fair)

Modified from http://www.iowadnr.gov/Environment/WaterQuality/WatershedImprovement/WatershedResearchData/WaterImprovementPlans/PublicMeetingsPlans.aspx. [Note that a new link to the Web page where the NFMR TMDL .pdf document is available under heading "Final Water Quality Improvement Plans."]

[a] Insufficient numbers of organisms for BMIBI calculation. To calculate the BMIBI, at least 1 of 3 quantitative benthic macroinvertebrate sample replicates must contain 85 or more individual specimens. The three replicates had 70, 25, and 54 specimens, respectively.

IDNR staff followed the Protocol for SI to determine the cause of the biological impairment in NFMR (see Tool # 3, *Stressor Identification (SI) and Causal Analysis/Diagnosis Decision Information System (CADDIS)*). The SI procedure relates impairments described by biological assessments to one or more specific causal agents (pollutants). It also separates water quality (pollutant) impacts from habitat alteration impacts. Although the SI did not reveal any single stressor that is clearly the dominant cause of biological impairment, IDNR determined that the primary pollutant-related causal factors in the NFMR were sediment, nutrients (specifically phosphorus), and ammonia.

Total Maximum Daily Load Development

In 2007, IDNR completed *Total Maximum Daily Loads for Sediment, Nutrients, and Ammonia: North Fork Maquoketa River, Dubuque County, Iowa*. Results of the SI process were used, and IDNR considered impacts from the point and nonpoint sources of pollution in development of the Total Maximum Daily Load (TMDL). Although IDNR concluded that one wastewater treatment plant in the NFMR watershed should be included in the TMDL and in developing a wasteload allocation for the existing phosphorus load, that facility did not contribute significantly to the overall sediment load.

IDNR also identified several potential nonpoint sources for nutrients, sediment, and ammonia—failed on-site septic tank treatment systems, agricultural activities (e.g., cattle in streams, fertilizer use, soil erosion, land-applied manure), wildlife, and runoff from developed areas (IDNR 2007).

To meet water quality improvement goals for the NFMR, the TMDL includes a 77 percent reduction in sediment reaching the river (20,200 pounds of sediment per year) and a 73 percent reduction in nutrients and manure releases. The TMDL has two parts. The first includes setting specific and quantifiable targets for sediment, oxygen demand, total phosphorus, and ammonia loads to the stream. Additional biological and water quality monitoring will determine whether the prescribed load reductions result in attainment of water quality standards. These monitoring data will also be used to determine whether the implemented TMDL and watershed management plans have been effective in addressing water quality impairments in the NFMR. EPA approved the IDNR TMDL in 2008.

IDNR has identified a variety of BMPs to improve water quality, as well as to encourage residents and businesses in the watershed to take action. IDNR has also identified possible practices to reduce sediment and nutrients reaching the NFMR, such as installing structures to reduce both agricultural and urban runoff; limiting cattle access to streams and installing alternative water sources for cattle; and using agricultural management practices that increase crop residue, such as no-till. IDNR also suggested that proper control of open agricultural animal feedlots will help prevent contaminated runoff from reaching streams, which in turn will reduce ammonia loading. Ongoing monitoring of this impaired stream segment will be used to periodically assess progress made toward attainment of the NFMR designated aquatic life uses.

3.8. Addressing Stormwater Flow in Connecticut's Eagleville Brook TMDL for Biological Impairment

Abstract

In 2004 Connecticut used biological assessment information to place Eagleville Brook on its 303(d) list of water quality limited (WQL) waters for failure to meet the brook's aquatic life uses. Before Total Maximum Daily Load (TMDL) development, the state conducted a stressor identification analysis that pointed to the complex array of pollutants transported by stormwater as the most likely cause of impairment.

A statewide study that correlated impervious cover (IC) with benthic macroinvertebrate data collected from wadeable streams was conducted, and results showed that the designated aquatic life use was not supported when IC was more than 12 percent of the watershed area. A TMDL was developed in 2007 using a target of 12 percent IC—the first in the nation to use IC as a surrogate for stormwater. Objectives to reduce IC were established for each waterbody segment, and progress toward attainment of the designated aquatic life use will be evaluated by monitoring the condition of the benthic macroinvertebrate community in conjunction with ongoing chemical assessments.

Eagleville Brook has a 2.4-square-mile drainage area, and the watershed drains a portion of the University of Connecticut (UCONN) campus and the town of Mansfield. The brook is designated as a Class A waterbody, but fisheries sampling in 2002 showed that the waterbody was not meeting its aquatic life uses, with low fish density and large areas with no fish. Additionally, benthic macroinvertebrate sampling in 2003 showed low total abundance and species diversity, documenting that the waterbody was in nonattainment of the state's narrative biological criteria for Class A waters. In 2004 Connecticut added Eagleville Brook to its list of impaired waters for cause unknown on the basis of the biological assessment results.

Stressor Identification and Total Maximum Daily Load Development

Before Total Maximum Daily Load (TMDL) development, Connecticut conducted a stressor identification (SI) analysis to evaluate the potential stressors and determine the most likely causes of impairment. The SI study concluded that biological impairments were most likely from a combination of pollutants related to stormwater runoff from developed areas and other related stressors (such as the physical impacts of stormwater flows). There are no other known point source discharges in this small watershed. The major source of stormwater is runoff from the impervious surfaces in the watershed (e.g., roads in Mansfield and UCONN campus). A statewide study of the impact of impervious cover (IC) on aquatic habitats was also conducted; Connecticut's Rapid Biological Assessment Protocol III data from 125 small (< 50 square miles) watersheds showed that no stream monitoring location with more than 12 percent IC in the upstream watershed meets Connecticut's biological criteria for full support of aquatic life use.

In 2007 Connecticut developed the TMDL with a loading capacity (TMDL target) of 12 percent IC. The 12 percent TMDL target was chosen on the basis of the threshold observed for applicable Connecticut streams in the statewide study. In the TMDL, Eagleville Brook was partitioned into three segments, and the IC was calculated for each. For each segment, a TMDL implementation objective was also developed (Table 3-7).

Table 3-7. Summary of TMDL analysis for Eagleville Brook

Waterbody segment	TMDL target	IC	Implementation objective
From the mouth at Eagleville Pond upstream to the confluence with Kings Brook, Mansfield	12%	5%	Antidegradation
The confluence with Kings Brook to headwaters near UCONN campus	12%	14%	21% reduction in the percent IC
Unnamed pond on UCONN campus	12%	27%	59% reduction in the percent IC

The targets apply at all times (instantaneously, daily, monthly, seasonally, and annually) and will achieve reductions in stormwater runoff volume in all storm events whenever they occur (e.g., on any day) throughout the year. The reductions associated with the implementation objectives were to be accomplished by improved stormwater management. The Connecticut Department of Environmental Protection (CT DEP) provided general and specific implementation recommendations in the TMDL and recommended using an adaptive management approach toward reducing stormwater impacts and improving water quality.

TMDL Implementation

Progress toward attainment of the aquatic life use will be evaluated by CT DEP's monitoring the macroinvertebrate and fish communities and assessing surface water chemistry according to an existing rotating basin sampling schedule. UCONN, the Town of Mansfield, and the Willimantic River Alliance have pledged support for TMDL implementation. EPA and CT DEP have funded a project using section 319 NPS funds to map locations and identify ways to reduce the effect of IC as required by the TMDL. The project also examined the estimated costs of such actions and developed initial engineering sketches for a *top ten* list for recommended retrofit management actions that are most cost-effective, primarily in the upper watershed. In addition, other projects have been completed on the UCONN campus to reduce IC, including installation of two green roofs and parking lots with pervious asphalt and concrete. The Town of Mansfield has received technical guidance on local land use regulations and practices, primarily in the lower watershed. Low-impact development concepts are expected to be incorporated into future development. An overall watershed management plan that supports a framework to pursue high-priority projects to reduce the effect of IC has been developed. Considerable stakeholder input has crafted a consensus approach to seize opportunities to reduce the effect of IC as situations arise during normal maintenance operations at UCONN and Mansfield. A tiered system to track progress will focus in the short term on close tracking of the area of new and disconnected IC, as well as flow monitoring to determine whether changes in IC will improve the hydrologic regime of Eagleville Brook. The TMDL has led to an increase in dialog among stakeholders and has led to changes in how people think about managing IC in the Eagleville Brook watershed. Additional information on the implementation of the Eagleville Brook TMDL can be found at http://clear.uconn.edu/projects /tmdl/index.htm.This site, hosted by UCONN, provides additional information and will be used to track the progress of TMDL implementation over time.

3.9. Vermont's Use of Biological Assessments to List Impaired Waters and to Support NPDES Permit Modification and Wastewater Treatment Facility Upgrades

Abstract

In the 1990s, the Vermont Department of Environmental Conservation's biological assessment of the Dog River showed aquatic life use impairments downstream of a wastewater treatment facility. Whole effluent toxicity and biological assessment data were used to support revisions to National Pollutant Discharge Elimination System permits for dischargers, and subsequent management actions at the facilities resulted in the segment's meeting its designated aquatic life use and its removal from the 303(d) listing for water impairment.

Biological Assessments Detect Impairment and Support Permit Modifications

Between 1993 and 1995, biological assessments of Vermont's Dog River showed that the river was not meeting its aquatic life use according to changes in the aquatic community typically associated with toxicity stress and moderate phosphorus pollution. In 1996, Vermont Department of Environmental Conservation (VT DEC) listed the Dog River on the state's 303(d) list of impaired waters, based on the biological assessment information, for cause unknown. Further investigation indicated two factors contributing to the degraded instream water quality. First, the Northfield Wastewater Treatment Facility (WWTF) had reached its design life and was no longer able to function properly and reliably meet National Pollutant Discharge Elimination System (NPDES) permit limits. Second, wastewater influent to the facility from two industrial textile facilities had high concentrations of metals and possibly surfactants. In WWTF effluent samples, metal concentrations were high and predicted to exceed water quality criteria at permitted flows. Whole effluent toxicity (WET) testing confirmed significant toxic effects at effluent concentrations greater than 12 percent. Through a toxicity identification evaluation (TIE) study, copper was identified as the most significant metal of concern in the WWTF effluent, with a maximum copper concentration of 184 micrograms per liter (μg/L). This level would have resulted in an instream concentration of 36 μg/L copper at 7Q10 (i.e., the lowest 7-day, consecutive low flow period occurring over the preceding 10-year period) permitted flows. Copper levels correlated with the level of toxicity found in the WET testing.

In 1999 pretreatment discharge permits with compliance schedules were issued to the textile facilities. The pretreatment permits established copper limitations for those influent waste streams that required the installation of pretreatment systems for the removal of copper (see Table 3-8). Although the systems were operational in 2000, biological assessments conducted between 2000 and 2003 showed continued aquatic life use impairment in the river. That monitoring showed a shift in the benthic macroinvertebrate community that, in addition to chemical data, indicated that phosphorus pollution had become the most likely cause of the aquatic life impairment. Specifically, the macroinvertebrate community was significantly higher in density and dominated by nutrient-tolerant taxa relative to previous sampling results. To measure this increase in nutrient-tolerant taxa, VT DEC used a ratio that compares the proportion of pollution- sensitive benthic macroinvertebrate species to more pollutant-tolerant species, the EPT/EPTc ratio. This reflects the ratio of generally pollution-sensitive species (e.g., **E**phemeroptera [mayflies], **P**lecoptera [stone flies] and **T**richoptera [caddisflies]) compared to the more pollutant-tolerant species (Chironomids [midges/flies]). A low threshold indicates dominance of midges (EPTc) that have been observed in streams with significant levels of nitrogen, phosphorus, or other pollutants.

Additionally, the higher biological index value reflected the increase in the midges and provides complementary information.

Table 3-8. Permit limitations for two textile facilities

Facility	Flow monthly average	Copper Monthly average	Copper Daily maximum
Facility A	150,000 gal/day	0.027 lb/day	0.038 lb/day
Facility B	35,000 gal/day	0.007 lb/day	0.0125 lb/day

In January 2003, VT DEC issued a compliance schedule to the Village of Northfield to upgrade its WWTF, and the upgraded facility became operational in November 2004. The upgraded WWTF process consists of upgraded headworks, two sequential batch reactors, a surge tank, and an upgraded chlorination and dechlorination system. Phosphorus removal was required to comply with the requirements of the Lake Champlain TMDL and Vermont regulations (10 VSA 1266a). To achieve that, permit limits for a 1.0-mg/day discharge of phosphorus were set at 6.78 lb/day, at concentration of 0.8 mg/L monthly average.

Northfield treatment plant copper effluent limitations were also established at 0.26 lb/day monthly average and 0.36 lb/max daily at a pH of between 6.5 and 8.5. Improved sludge management was also incorporated into the upgraded WWTF, including refurbishing the existing digester, adding a new digester, and adding a centrifuge for dewatering. Water quality and habitat improvements were observed, but the aquatic system's recovery was further complicated by a chlorine spill from the WWTF's temporary disinfection system during the upgrade in July 2004, leading to a further short-term decline in EPT.

Conclusion

Despite the short-term adverse effects from the 2004 chlorine spill, the compliance schedules and changes to both predischarge and the WWTF permits have resulted in changes in facility operations that, in turn, have resulted in improvements in water quality. Biological assessments showed improvement only after copper was reduced and wastewater treatment of phosphorus was improved.

These combined efforts enabled a site that was classified as fair-poor to recover to excellent condition. Biological assessments in 2005 and 2006 showed that the Dog River was meeting its aquatic life uses, with specific measures, or metrics, showing density to be moderate; richness, EPT, and EPT/EPTc ratio to be high; and biological index (BI) to be lower relative to previous sampling.

Chemical monitoring has documented that the applicable chemical water quality criteria were being met, and WET test results have shown that the effluent is nontoxic (i.e., no significant toxicity to test organisms using 100 percent effluent). The biological assessment information documents that the stream macroinvertebrate community is now dominated by water-quality-sensitive taxa more typical of its *natural* expectation— with recovery of sensitive species and a more balanced community. (Data from sampling between 1993 and 2006 are shown in Table 3-9.) As a result, in 2006 Vermont removed Dog River from its impaired waters list.

Table 3-9. Macroinvertebrate assessments for Dog River—Northfield WWTF

Date	1993	1994	1995	2000	2001	2003	2004	2005	2006
Assess (criteria)	Fair	Fair	Fair	Fair-Poor	Poor	Poor	Poor	Very Good	Excellent
Density (> 300)	1,862	3,282	1,037	4,556	5,640	4,264	668	2,160	5,870
Richness (> 30)	39	43	41	50	50	62	34	51	62
EPT (> 18)	12	16	16	14	11	22	12	28	33
BI (< 5.00)	4.73	4.74	4.61	5.51	6.00	5.26	5.12	4.38	3.48
Ept/EptC (> 0.45)	0.029	0.50	0.52	0.29	0.07	0.22	0.14	0.89	0.89

Milestones:

2000 – Metals removed.

2004 – Chlorine spill late summer; WWTF upgrade with phosphorus removal completed in November.

2005 – First year of river meeting designated aquatic life use.

2006 – Second year of river meeting aquatic life use; stream removed from impaired waters listing.

3.10. Restoration of Red Rock Creek by the Grand Portage Band of Lake Superior Chippewa

Abstract

For the past 15 years, the Grand Portage Band of Lake Superior Chippewa (tribe) has led efforts to restore one of the Band's most impaired waters—Red Rock Creek. Biological assessment information has played a central role in establishing and assessing whether biological, chemical, and physical targets for restoration are being met. To date, the tribe has implemented multiple and interrelated restoration activities that have resulted in significant water quality improvements, as demonstrated by periodic sampling of the creek's benthic macroinvertebrate and plant communities.

Background

Over the past decade, the Grand Portage Band of Lake Superior Chippewa (tribe) has been leading restoration efforts to improve the physical, chemical, and biological integrity of one of the Band's impaired waters—Red Rock Creek. To date, biological assessment information has played a central role in defining biological goals for restoration in concert with chemical and physical targets that have also been established. The tribe has implemented restoration activities that have resulted in water quality improvements, as shown in sampling of both the benthic macroinvertebrate and plant communities.

Red Rock Creek Impairment

The Red Rock Creek watershed encompasses approximately 1,200 acres in Minnesota north of Lake Superior. While the upper reaches of the watershed are in relatively pristine condition, the creek flows through an abandoned gravel pit located approximately one-half mile from Lake Superior. Past gravel mining activities—most notably the removal of riparian (streamside) vegetation and cutting of a portion of the stream bank—have adversely affected the stream, resulting in severe sedimentation. This has resulted in a net loss of fish species and benthic macroinvertebrate communities. For instance, by 2006, steelhead trout, chinook salmon, coho salmon, and coaster brook trout were found only near the mouth of the stream, rather than their previous habitation along several miles of the stream. Gravel extraction has also caused the stream to leave its former channel and to spread into the gravel pit area. Notably, beaver damming has exacerbated problems associated with braiding and flow and has led to clogging of Red Rock Creek.

Monthly sampling of Red Rock Creek began in 1997. Turbidity measurements were high, with a mean concentration of 12.3 nephelometric turbidity units (NTUs). Gravel mining activities ceased in 1998, and in 2000 the Tribe reported that water quality was impaired based on biological and chemical assessments. Specifically, monitoring showed low dissolved oxygen concentrations, high turbidity, and low benthic macroinvertebrate densities and species abundance. In the impacted portion of the creek, mean dissolved oxygen concentrations were 6.3 mg/L—more than 2 mg/L lower than the concentrations measured in unimpacted upstream reaches. A total of27 macroinvertebrates were collected in the impacted stream reach, with a large proportion of pollution/sediment-tolerant diptera (e.g., Chironomides [midges]) present but no pollution-sensitive EPT taxa (e.g., **E**phemeroptera [mayflies], **P**lecoptera [stone flies] and **T**richoptera [caddisflies])). However, in 2004, 6 years after the cessation of gravel mining operations, over 100 macroinvertebrates were collected. Possible explanations for this improvement in macroinvertebrate density might be the subsequent regrowth of some of the stream's riparian buffer and instream habitat (Table 3-10). However, only 27 percent of the total taxa were EPT taxa, which is much lower than the 60–75 percent proportion of EPT taxa expected in unimpacted or minimally impacted streams in this area. Increases in EPT taxa are expected with continued restoration and allowing time for the aquatic system to recover natural flow and habitat conditions.

In addition to benthic macroinvertebrates, the tribe also assesses plant communities to evaluate the biological health of its waterways. To measure the natural quality of the area, the tribe uses a Floristic Quality Index (FQI),[17] a weighted species richness index that can be calculated by identifying all plant species in a given plot or transect. To evaluate streams, the Grand Portage Tribe uses an FQI score ≥ 20, the presence of at least 20 plant taxa, no exotic invasive plant species, and at least 5 sensitive or rare plant taxa. In 2004, Red Rock Creek had a total of 13 plant taxa, an FQI score of 14, 3 invasive exotic plant species, and no sensitive or rare plant taxa (Table 3-11).

Restoration Efforts

The tribe set biological, chemical, and physical goals for improving overall water quality in Red Rock Creek (Table 3-10). Restoration goals were established for increased dissolved oxygen concentrations, reduced turbidity, reduced diptera taxa to less than 5 percent of macroinvertebrates collected, and increased proportion of pollution-sensitive macroinvertebrate taxa. Restoration efforts began in 2006 with the removal of the beaver dam and installation of sediment traps. Monitoring results conducted immediately following restoration showed a mean turbidity concentration of 10.3 NTUs, dissolved oxygen concentrations that continued to be approximately 2 mg/L less than those in undisturbed reaches of the stream, and changes in the benthic macroinvertebrate community. Although sampling of the macroinvertebrate community showed a dramatic increase in the number of organisms collected (350), only 9.8 percent of the total insects collected were EPT taxa and 22 percent were diptera—similar to pre-restoration sampling results. In 2008 additional restoration measures were completed, including reinforcement of banks upstream of the sediment basin using live fascines and stakes, physical removal of excess sediment from the basin, and seeding and tree planting to further stabilize the banks and restore riparian vegetation.

Results

Monitoring results from 2008 and 2009 show that the restoration goals for Red Rock Creek have been exceeded for most biological, chemical, and physical measures of water quality (Tables 3-10 and 3-11). Dissolved oxygen concentrations and turbidity levels are comparable to those expected in unimpacted conditions with improvements in both benthic and floristic assessments of biological condition, though the continued presence of invasive plant species remains a challenge. The tribe will continue to maintain the sediment ponds and bank stabilization projects in order to achieve the restoration goal for percent EPT taxa. Regular removal of excess sediment from the basin, efforts to reestablish native vegetation in the riparian zone, and potential removal of invasive species from the basin will be considered in an adaptive management approach to fully achieve biological restoration goals.

Table 3-10. Sampling to assess progress toward restoration goals

Parameter	Pre-restoration sampling results (year)	Restoration goal	Post-restoration sampling results (year)
Turbidity	12.3 NTU (1997)	50% reduction	2.4 NTU (2009)
Dissolved oxygen	6.3 mg/L (2000)	2 mg/L increase	9.6 mg/L (2009)
Number of macroinvertebrates	27 (2000) 10 (2004)	200	350 (2008)
% diptera	29.6% (2004)	Reduction to 5% of total	1.3% (2008)
% EPT species	27% (2004)	Increase to 60% of total	30% (2009)

Table 3-11. Plant sampling results

Parameter	2004	2008
Number of plant taxa	13	21
FQI score	14	19
Number of invasive plant species	3	3
Number of sensitive or rare taxa	0	3

3.11. Using Biological Assessment Data to Show Impact of NPS Controls in Michigan

Abstract

In the 1990s biological assessments of Carrier Creek in Eaton County, Michigan, showed that the waterbody was not attaining its designated aquatic life uses, resulting in its inclusion on the state's 303(d) list in 1996 for cause unknown. Subsequent surveys indicated that stream biota was affected by urban runoff, poor instream habitat, and sediment deposition. In 2002 a Total Maximum Daily Load for biota was completed. Watershed partners are conducting several stream restoration projects to improve aquatic life use attainment. The restoration activities stabilized the stream channel and its hydrology, reduced stream bank erosion, and improved aquatic habitat. Improvements in fish and macroinvertebrate communities have been documented.

Background

Carrier Creek, a tributary to the Grand River, flows through a rapidly developing area in Eaton County near Lansing, Michigan. Historical channelization and more recent urban runoff resulted in eroding stream banks, high sedimentation rates, and degraded aquatic habitat for fish and macroinvertebrate communities. In 1996 Michigan included a 4-mile segment of the creek—from its confluence with the Grand River upstream to where it flows under Interstate 496—on its 303(d) list of impaired waters based on biological assessment information used to interpret its narrative standard that all surface waters of the state are "designated for and shall be protected for ... aquatic life and wildlife." The Michigan Department of Environmental Quality (MDEQ) determined that the quality of the aquatic biota in that segment of the creek was reduced by urban runoff, poor instream habitat, and excessive sediment deposition. MDEQ completed a Total Maximum Daily Load (TMDL) for Carrier Creek biota in 2002. As noted in the TMDL, achievement of the water quality standards (WQS) for designated uses for Carrier Creek will be demonstrated by assessing the macroinvertebrate community and the instream habitat as it relates to sediment.

Stream Restoration

Between 2000 and 2006, state and local agencies and volunteer groups partnered in various stream restoration projects designed to achieve the TMDL goals. For example, in 2000 local agencies and volunteers stabilized and restored 5 miles of channel. The projects increased channel stability, improved instream habitat, and reconnected the channel to its floodplain. The upstream end of the channel was narrowed, and the stream pattern was reestablished to promote meandering. In some locations, the project team removed dredge spoils that were separating the stream from its natural floodplain.

In 2002 project partners created a 32-acre wetland in the headwaters of the watershed to intercept stormwater runoff and decrease stream flashiness. In 2004 the Perrin Section of Trout Unlimited installed structures along the creek to provide shelter and resting points for fish. In addition, the Eaton County Drain Commissioner is enhancing stormwater detention and flow control throughout the upper portion of the watershed to stabilize the channel, reduce the velocity of the flow, reduce erosion downstream, and reduce the amount of flooding. That work is ongoing.

Results

Biological assessment data have been used to assess the project's progress. The State of Michigan and the Eaton County Drain Commission collected data on fish, macroinvertebrates, and aquatic habitat quality at two locations in the project area, both before (2000) and after (2006) the restoration activities occurred. A consultant for the Eaton County Drain Commission collected additional fish data in 2007.

As of 2006, aquatic habitat was unchanged at one site and had improved at the other, but macroinvertebrate populations had not responded. However, by 2009, both macroinvertebrate and habitat quality scores had improved at all sites. The improvement in habitat scores was due to continued stream restoration activities that provided meandering channels and suitable instream habitat for the aquatic biota, such as fish and benthic macroinvertebrates. In fact, the 2007 fish data show that the number of fish taxa increased at both locations following restoration activities, more than doubling at one site and quadrupling at the other. There is another encouraging signal of improvement to date: a single slippershell mussel (*Alasmidonta viridis*) was found during an informal inspection of the restored reach in 2007. The slippershell is listed on the state's threatened list by the Michigan Natural Features Inventory and had not been observed in the stream before restoration. MDEQ will conduct further monitoring in the fall of 2011.

The restoration activities conducted to date have stabilized the stream channel and its hydrology, reduced stream bank erosion, and improved aquatic habitat. Fish communities are recovering, and future monitoring should show further improvements in the biota and eventually result in removing Carrier Creek from the list of impaired waters based on assessing the macroinvertebrate community and the instream habitat as it relates to sediment.

3.12. Using Biological Assessment as Evidence of Damage and Recovery Following a Pesticide Spill in Maryland and the District of Columbia

Abstract

In response to a fish kill in a tributary of the Potomac River in 2000, biological assessment data were used to show the impact of a pesticide spill and to document the waterbody's recovery. Sampling data collected before the spill provided a baseline of the expected aquatic community in the waterbody. Data from biological assessments before the spill were compared with sampling data collected immediately after the fish kill and several months later. The data were used to support enforcement actions and to support criminal charges against the polluter.

Problem Overview

In the spring of 2000 a fish kill (estimated to be 150,000 fish) was observed along an 8-mile stretch of Rock Creek, a major tributary of the Potomac River in Maryland and the District of Columbia.

Responding to the kill, the Maryland Department of the Environment (MDE) sampled the water column and sediments and found high concentrations of the insecticides cypermethrin and bifenthrin, both of which are highly toxic to fish. Concentrations were especially high in a storm drain entering the stream from the parking lot of a pest control company, suggesting that a pesticide spill had occurred.

The case was investigated by EPA's Criminal Investigation Division with assistance from the State of Maryland, Montgomery County, the National Park Service, and the District of Columbia. Within 2 weeks, a coordinated, multiagency effort sampled sediments, fish, and benthic macroinvertebrates upstream and downstream of the outfall. Fish sampling was repeated after 5 months, and sediments were retested 9 months after the spill.

Data Collection and Analysis

Samples were analyzed in three time frames—before the spill occurred, just after the fish kill was observed, and some months afterward. Samples were also categorized by location; before and upstream samples served as controls for the suspected effects of the spill. Several hours after the fish kill was first observed, cypermethrin and bifenthrin concentrations in downstream waters were near the acute toxicity thresholds for fish and invertebrates. Pesticide concentrations in the storm drain were many times greater than the acute toxicity levels. Sediments tested 2 weeks after the fish kill showed elevated levels of cypermethrin and bifenthrin below the storm drain when compared to levels above the storm drain. When retested 9 months later, cypermethrin and bifenthrin concentrations in all sediment samples were below detection limits.

Fish and benthic macroinvertebrates were collected from 11 stations, including 4 above and 7 below the storm drain. Several sites had been sampled before the spill in routine monitoring programs by the District of Columbia and Montgomery County. Historical data from 1996–1998 were available for three stations below the outfall, and one site well below the spill had been sampled several times weeks before the spill. Just after the spill, both fish and macroinvertebrate communities showed severe degradation when compared to upstream controls and, for fish only, when compared to downstream samples taken before the kill event.

Decreases in numbers of fish and the number of fish species were observed, with a reduction in the fish index of biotic integrity at all sites below the spill. On average, 20 macroinvertebrate taxa, of 46 taxa found upstream, were absent from downstream sites. After 5 months, most minnow species had returned to the affected sites. Overall, the fish community had recovered to approximately 75 percent of upstream species composition.

Conclusion

Biological assessment provided a powerful tool for documenting stream degradation and stream recovery following the toxic spill. Evidence was further strengthened by baseline data collected in routine monitoring programs. Comparison of the post-spill samples to samples taken before the spill provided a quantitative assessment of the biological impact and evidence of stream recovery. In November 2001, the owner and an employee of the pest control firm were charged with violations of the Clean Water Act and the Federal Insecticide, Fungicide, and Rodenticide Act. Ongoing biological assessments, in conjunction with bioassays and chemical and physical assessments, can assist enforcement agencies in assessing damage, levying fair and reasonable damage assessments on those proven responsible for toxic spills, and determining the rate and level of stream recovery.

3.13. Support for Dredge and Fill Permitting in Ohio

Abstract

Ohio uses biological assessments to help inform its decisions about certifying permits for dredge and fill activities and to ensure that the impacts of those activities on aquatic habitats do not violate Ohio water quality standards (WQS). Ohio's tiered aquatic life uses, in conjunction with antidegradation policies and numeric biological criteria adopted into the state's WQS, enable Ohio to better assess the potential impact of dredge and fill activities and to make management decisions on the basis of its designated aquatic life uses. Ohio's designated aquatic life uses are based on the relationship of habitat and the resident biota.

It is presumed that if critical aquatic habitat is present or can be restored, the aquatic life associated with the habitat can be supported. Additionally, when implementing nationwide permits, Ohio has been able to include additional conditions to protect high-quality waters as revealed by biological assessments.

Dredge and Fill Permitting

States use Clean Water Act (CWA) section 401 to regulate activities that might impact aquatic habitats. Those wanting to modify a stream in a way that will result in the discharge of dredge or fill material into waters of the United States must obtain a section 404 permit from the U.S. Army Corps of

Engineers and a section 401 water quality certification from the state. The state must certify that the proposed activities will comply with and not violate water quality standards (WQS) or waive such certification. Ohio's designated aquatic life use classes, which are based on the relationship of habitat and the attendant numeric biological criteria adopted into the WQS, make that linkage a valid tool for evaluating the effects of habitat alterations that are covered under the CWA. In essence, the habitat tools employed are sufficiently predictive to serve the purpose of reviewing proposed stream habitat modification activities.

Ohio EPA used more than 20 years of data to develop habitat stressor gradients along several aspects of habitat quality at both the site and watershed scales, including overall habitat quality as measured by a habitat quality index, the Qualitative Habitat Evaluation Index (QHEI), and for specific attributes such as substrate and channel condition (Rankin 1989, 1995). This allows for sufficient predictive relationships such that this habitat tool can be used to help determine the attainability of the Ohio biological criteria.

Ohio's designated aquatic life uses for surface waters have enabled a range of management responses to dredge and fill projects related to the quality and sensitivity of the waterbody in the context of the CWA goal to protect aquatic life. Ohio's use classification system is tiered along a gradient of quality with the highest use class supporting pollution-sensitive, naturally occurring communities of benthic macroinvertebrates and fish (Exceptional Warmwater Habitat [EWH] Aquatic Life Use). A second class along the gradient (Warmwater Habitat [WWH]) also supports a community of pollution-sensitive, naturally occurring benthic macroinvertebrates and fish species that are consistent with least impacted reference conditions.

Nationwide permits are designed to minimize site-specific oversight where ecological risks are assumed to be low. Frequently, however, in reviewing the criteria where nationwide permits can apply, high-quality waters can be overlooked, leading to their unwarranted alteration and impairment. Small streams such as headwater streams are particularly vulnerable to not being properly assessed under nationwide permit conditions. The Ohio EWH use designation requires high-quality habitat and stable hydrological regimes (especially in headwater and wadeable streams). Because those essential attributes can be altered by direct modifications to the stream channel and other habitat features, Ohio requires individual reviews of projects that occur in such high-quality streams. Under a general use system, those sites would be lumped with all other streams under the nationwide permit system. In addition, antidegradation provisions for high-quality WWH and Coldwater Habitat (CWH) streams are also applied.

Mitigation Standards

The attention gained by biologically defined habitat impacts has prompted the development of mitigation standards, in conjunction with a process for rigorous validation, that will take Ohio's aquatic life uses into account and require enhancement or restoration wherever feasible. The stressor-response relationships that have been developed between biological assemblages and key habitat attributes have been applied to the 401 program in Ohio for more than 20 years. For nationwide permits, a series of general and specific exclusions and conditions that vary with the state's tiered uses have been derived (USACE 2002). They include a general exclusion (of nationwide permits) for streams that are EWH and for certain high-quality antidegradation tiers (State Resource Waters and Outstanding State Resource Waters, Superior High-Quality Waters), the delineation of which was based primarily on the same biological assemblage attributes on which the designated use classes are based.

Ohio's integrated approach for designating aquatic life uses, implementing antidegradation, and establishing biological criteria is based on relationships between the aquatic biota and critical aquatic habitat.

3.14. Virginia INSTAR Model for Watershed Protection

Abstract

The Virginia Department of Conservation and Recreation and Virginia Commonwealth University Center for Environmental Studies are collaborating in developing and implementing a statewide Healthy Waters program to identify and protect healthy streams.

The Interactive Stream Assessment Resource (INSTAR) is an online, interactive database application that evaluates the ecological integrity of Virginia's streams using biological and habitat data. The Web-mapping application is available to the public as a free resource to help planners, advocacy groups, and individuals to support wise land use decision making.

In 2003 Virginia Commonwealth University's Center for Environmental Studies, Virginia Department of Conservation and Recreation (VA DCR), the Virginia Department of Environmental Quality, Virginia Coastal Zone Management Program, and other state agencies began collaboration on Interactive Stream Assessment Resource (INSTAR). INSTAR is an online, interactive database application that evaluates the ecological integrity of Virginia's streams using biological assessments and habitat data. INSTAR was developed as part of and to support Virginia's Healthy Waters Initiative. That initiative is an effort to raise awareness of the importance of stream ecological condition and how healthy it is and to make certain that conservation efforts are broad enough to include healthy streams and rivers, making them and restoration efforts a priority. The approach is complementary to water quality programs that focus on repairing degraded streams.

INSTAR is used to identify healthy streams using data that include information about fish and macroinvertebrates, instream habitat, and riparian borders. Users can access and manipulate the view of a comprehensive database representing more than 2,000 aquatic (stream and river) collections statewide. INSTAR was established to develop complementary, synoptic, and geospatial database for fish and macroinvertebrate community composition and abundance at stream locations throughout the state. INSTAR, and the extensive aquatic resources database on which it runs, supports a wide variety of stream assessment, management, and conservation activities aimed at restoring and protecting aquatic living resources throughout Virginia.

INSTAR was primarily designed as a tool that could be used for regional and local planning by providing support for making land use decisions and help in prioritizing stream protection and mitigation efforts. Advocacy groups and individuals might also want to use INSTAR to identify healthy streams in their communities and encourage their protection. INSTAR can support regional approaches to transportation, priority habitat corridor identification, greenways, zoning, and land conservation priorities. It can also be used to identify healthy streams vulnerable to development and those already protected. Locally, INSTAR can help raise awareness about the location of healthy waters and identify priority areas during comprehensive planning. Measures of the composition of the naturally expected benthic macroinvertebrate community provide a benchmark for determining a healthy stream.

INSTAR generates a Virtual Stream Assessment (VSA) score for each stream studied using data collected by biologists along a 150- to 500-meter length or reach of stream, depending on its width. Information collected includes the types and number of fish and aquatic macroinvertebrates, instream habitat (e.g., vegetation, rocks, fallen logs), and riparian vegetation. The information is compared statistically to a model reference stream that represents ideal conditions of biology and habitat for streams in that geographic region. How closely a stream compares to an appropriate model reference stream determines its VSA score and ranking. That information can help identify a range of condition, from streams that have exceptional health to streams that are good candidates for restoration. INSTAR also classifies Virginia's 1,275 small watersheds using a modified index of biological integrity (mIBI) that is based on occurrences of selected aquatic species found in each watershed.

With INSTAR, a user can generate stream data and mapping information at the local, regional, or statewide level. Searches can be done by locality, stream name, watershed, or drainage area, and specific locations can be pinpointed using global positioning system (GPS) coordinates or street addresses. Users can also access information about fish, macroinvertebrates, and habitat for a specific stream location and can turn on topographical views, road maps, wetland overlays, and aerial photos. Users can also measure, outline, and highlight areas; add and edit text; and generate customized maps and reports. INSTAR is available to the public through a free, user-friendly website: http://instar.vcu.edu.

Application of INSTAR in Richmond County

The Richmond County Local Tributary Strategy Pilot Project, funded through grants from the National Fish and Wildlife Foundation and VA DCR, focused on the capacity of stakeholders to develop and support a local program to implement statewide strategies to mitigate nutrient and sediment pollution delivered to local waters and the Chesapeake Bay. The project approach identified aspects of local/regional planning and implementation programs where consideration of strategies to meet regional water quality goals could lead to improved condition or improved protection of natural resources. The best outcome would be that implementation would affect local needs and the broader Chesapeake Bay goals. County-comprehensive planning and agricultural best management practice (BMP) implementation programs are examples of local programs that vary greatly in how they are managed and have regional impact. Central to success in the project was identifying a way to link such varied efforts so that their strategies might align with regional goals. The project worked to establish that link through a focus on linking land use to water quality or stream health. The link was defined by two data-collection efforts. A countywide INSTAR stream assessment was conducted, and a countywide chemical water assessment was conducted.

The stream health assessment became a central theme for the project as the data were reviewed under several different contexts.

1. The project participated in the county-comprehensive plan review and revision process as a partner in an extensive community engagement process. Work sessions were held to specifically discuss the link among land use, management and planning, stream health and natural resource conditions and trends, and a host of other social and economic sector interests. The stream health assessment was an important component of the natural resource workshop.
2. INSTAR-identified healthy stream sites were included as a component of secondary considerations in the local Soil and Water Conservation District Agricultural BMP Cost Share Program guidance.
3. The INSTAR stream assessment was used in combination with the chemical water quality and agricultural BMP implementation data to correlate stream health and the level of BMP implementation or the percentage of land treated in a site's drainage area. The map displays an enhanced view of INSTAR data that includes sites identified as Important Fisheries Resources and their spatial distribution against the level of BMP implementation in corresponding watersheds.
4. The INSTAR stream assessment was used to review the health of streams that received drainage from the main urbanized area affecting the county's jurisdiction. The data allowed for prioritizing sites where improved stormwater management could affect local conditions and regional implementation goals.
5. The comprehensive nature of the stream assessment provides a baseline condition for the local effort to measure progress, impacts, identify threats, or conservation priorities.

The regional strategies developed under Virginia's initial Tributary Strategies and revisited in the development of the Chesapeake Bay Total Maximum Daily Load (TMDL) do not provide local data to assist with implementation planning. The INSTAR stream assessment is a way to fill that data gap.

3.15. Examination of Climate Change Trends in Utah

Abstract

U.S. Environmental Protection Agency and the Utah Department of Environmental Quality (UT DEQ) are partnering in analysis of long-term biological assessment data to evaluate the potential impact of global climatic trends on the aquatic biota in Utah's streams. UT DEQ's objective is to develop a defensible approach to account for systematic bias that these impacts might have on its biological assessment and biological criteria program. Reference condition (e.g., natural or near natural condition) provides a baseline for comparison between expected conditions and test sites so it is important for states to understand and, where possible, quantify the shifts in the *steady state* of local reference communities due to global climatic shifts, regardless of whether they are natural or human-induced. For example, test sites should not be expected to exhibit communities that no longer exist at reference sites. UT DEQ's objective is to quantify the proportion of variation attributed to temperature-driven effects.

U.S. Environmental Protection Agency (EPA) Office of Research and Development (2010c) analyzed biological assessment data from Utah to determine whether past climate trends could be detected and to characterize the vulnerabilities of the biological assessment program to future climate conditions. In particular, the Utah Department of Environmental Quality (UT DEQ) was concerned that systematic changes in the physical or biological characteristics of streams would bias biological assessment scores, leading to errors in its integrated report. The availability of long-term stream invertebrate data at four reference stations, in two ecoregions, formed the basis for the analyses.

Long-term declines in richness or abundance of cold-preference taxa was detectable (i.e., from statistically significant temporal trends) at the two longest-term (> 15 years) Utah reference stations— one in the Wasatch-Uinta ecoregion and the other in the Colorado Plateau. That response was supported by significant associations between declining richness or abundance of cold-preference taxa and increasing temperature. Fairly predictable losses in a metric considered sensitive to pollution and disturbance, EPT taxa richness, were observed with increasing temperatures at the locations, which represent both high- and low-elevation ecoregions. The EPT metric is a measure of the presence of generally pollution-sensitive species (e.g., Ephemeroptera (mayflies), Plecoptera (stone flies) and Trichoptera (caddisflies)) in a sample. The response of EPT taxa was largely driven by losses of coldwater-preference EPT taxa, but in some cases it was also influenced by gains in warm-preference EPT taxa.

From those results, it was estimated that a 25 - 40 percent loss of EPT taxa could occur with current scenarios of temperature increases by 2050 (USEPA 2010c). Should such substantial losses of EPT taxa due to climate change occur, it would confound measures of ecological condition and decisions regarding attainment of aquatic life uses in many state monitoring programs. The Utah results suggest that relative elevation is a contributing factor driving the temperature trait composition of regional benthic communities (USEPA 2010c), with a greater proportion of cold-preference taxa in the higher elevation ecoregions and a greater proportion of warm-preference taxa in low-elevation ecoregions. Higher elevation regions with a greater proportion of cold-preference taxa might have a greater vulnerability to temperature-driven effects on traditional, taxonomically based indicators of biological condition. However, with the results of these studies and others, temperature-modified metrics can be used to characterize the contribution of climate changes in temperature to the observed trends, which would minimize both false-positive and false-negative decisions about aquatic life use support.

UT DEQ uses a mathematical model, River Invertebrate Prediction and Classification System (RIVPACS), to predict the expected composition of benthic macroinvertebrate species inhabiting streams from observations made at numerous streams that are relatively unimpacted by anthropogenic stress. The expected composition provides the baseline against which a test stream is compared. The results of the study show that changes in climate-related parameters used as predictor variables in the model will potentially alter the model's precision. The model needs to be calibrated for the climate-sensitive parameters so that effects from global climate change (regardless of whether they are natural or enhanced by anthropogenic sources of carbon to the atmosphere) and effects from anthropogenic stress (e.g., toxic discharges, stormwater flows, nutrient enrichment) can be distinguished. UT DEQ recalibrates the model every 2 years for Integrated Report purposes. Recalibration includes new reference sites, updated data from existing reference sites, and new environmental predictor variables and data. Therefore, as part of its existing program, Utah is able to accommodate and adjust for changes to predictor variables due to climate change, provided that it is aware of the potential for systematic bias.

To continue support of the effort, UT DEQ intends to collect additional data at long-term reference sites. Using the initial 2006 RIVPACS model as baseline, which includes most historical data from reference sites, at least five sites from each of the eight biologically similar groups will be sampled. A site will be sampled when the basin rotation monitoring plan is implemented for that basin (six-basin rotation). The sites encompass various levels of elevation, watershed size, latitude, and such, which can provide clues where climate-change effects are most pronounced. The RIVPACS model will be recalibrated every 2 years including new reference sites and updated predictor variable data. These recalibrated models will then be applied to data collected from the revisited trend reference sites to quantify several measures of long-term biological changes, including observed/expected (O/E) trends sites, changes in biological group membership, and taxon-level changes within group membership, including patterns in trait-based community composition. Site-specific results from these recalibrated models will also be compared to historical results to evaluate the extent to which climate trends would have altered decisions regarding support or non-support of aquatic life uses if climate-related biases were not accounted for in the analyses.

3.16. Applications of Biological Assessment at Multiple Scales in Coral Reef, Estuarine, and Coastal Programs

Abstract

Biological assessments provide useful information on the cumulative impacts of multiple stressors on biological conditions. As integrators, biological assessments can also evaluate the effects of landscape and ecological processes on aquatic life. By applying biological assessments at multiple spatial scales and multiple levels of biological organization in large and spatially complex waterbodies such as estuaries, coral reefs, or large braided river networks, U.S. Environmental Protection Agency hopes to expand its ability to understand first the interactions of biological communities with the large-scale processes that define ecosystems and second the cumulative effects of multiple stressors over larger spatial scales and over decadal time periods. Approaches combining biological assessments at several scales and levels are being developed for estuaries in the National Estuary Program and for coral reefs.

Background

Biological assessments can be conducted at many spatial scales and at many levels of biological organization. *Spatial scale* refers to the area considered in a biological assessment and can range from a shoreline or stream reach to an entire waterbody, region, state, or nation. Level of biological organization makes note that biology self-organizes into levels of order or structure such as organism, population, community, biotope, bioregion, or biome. Each level is generally associated with a physical space, such as habitat, landscape, watershed, or region. For example, biological assessment is a valuable tool to examine a single stream reach by considering the biological community within a defined habitat or a consolidated group of habitats in the stream (USEPA 1990, 1999). Such habitat-specific community-level biological assessments can also be conducted at local, state, and national spatial scales. U.S. Environmental Protection Agency's (EPA's) National Coastal Assessments (2001–2006) and National Coastal Condition Assessment (2010)[18]—programs designed to assess the condition of the nation's estuaries and coastal waters—conduct habitat-specific community-level biological assessments (hereafter referred to as habitat-level assessments) at the national scale. Habitat-level assessments are consistent with the definition of biological integrity as the capability of supporting and maintaining a balanced, integrated, adaptive community of organisms having a composition and diversity comparable to those of natural habitats of the region (Frey 1975; modified by Karr and Dudley 1981).

At a different level of biological organization, several methods for biological assessment that are specific to the aquatic landscape or to landscape-level processes have been developed. These methods can be useful tools in spatially complex waterbodies that are defined by interconnections among biological communities and among many distinct environments or habitats. Landscape-level concepts can be applied to all waterbody types and provide particular insights for watershed management. They are potentially very helpful as evaluative tools in waterbodies that appear as intertwined, patchy (and often shifting) mosaics of environments that support different biota and respond differently to different stressors.

Coral Reef Biological Assessments

The concept of biological integrity at the landscape level has, for example, been identified as important in developing biological criteria for coral reefs. Coral reefs are spatially complex habitats that are inextricably intertwined with a larger set of adjacent habitats (e.g., mangroves and seagrasses). Coral reef biota have evolved life history strategies that rely on the availability of those adjacent habitats (Christensen et al. 2003; Mumby et al. 2004, 2008; Aguilar-Perera and Appeldoorn 2007; McField and Kramer 2007; Meynecke et al. 2008; Sale et al. 2008). EPA's Coral Reef Biological Criteria document (Bradley et al. 2010) points out that "[b]iological integrity also means that reef organisms...have a clean, healthy environment to support them, including habitats for propagation, nurseries, and refugia. In this context, a fully functioning coral reef ecosystem may include adjacent supporting ecosystems such as seagrasses and mangroves." That document also recommends area measures of coral reef extent (e.g., square meters) as a first-order method for biological assessment of coral reefs that is relevant to landscape-scale evaluations. While most monitoring programs portray coral quantity as two-dimensional (2-D) live coral cover, EPA has developed a rapid survey procedure for estimating three-dimensional (3-D) total coral cover, which more realistically characterizes coral structure available as community habitat (Fisher 2007; Fisher, Davis, et al. 2007; Fisher, Fore, et al. 2008).

In conjunction with National Oceanic and Atmospheric Administration (NOAA) and other partners, scientists from EPA's Atlantic and Gulf Ecology Divisions (Narragansett and Gulf Breeze) are exploring the use of biological assessments to describe the coral reef and fish community along a gradient of stress in Guánica Bay, Puerto Rico. This effort may expand to include other critical coastal habitats in the future, e.g., sea grass beds and mangrove forests. Scientists will examine the pollution sensitivity of different taxa, presence or absence of native species, and other ecological response variables and then map the changes in these variables along a gradient of increasing stress—a Biological Condition Gradient (BCG) (see Section 2, Tool #2). Additionally, if there is sufficient quality and quantity of field data available, the BCG can provide a framework for relating well documented numeric stressor-response relationships to biological condition and thereby more precisely define stressor concentrations that support a waterbody's designated aquatic life use. Establishing this relationship could involve two steps. One step is establishing a numeric biological threshold that corresponds to the desired level of biological condition. For example, State and Tribal programs often develop numeric biological thresholds based on reference site conditions using an index of biotic integrity (IBI) or modeling the ratio of observed to expected species (O/E). Quantifying the relationship between BCG tier assignments and IBI or O/E scores for sampling sites along a gradient of stress provides a mechanism to link the scores to different levels of biological condition. The other step is quantifying the relationship between the IBI or O/E values and the stressor/parameter of interest such as nitrogen or phosphorus. Once a significant relationship between the IBI or O/E values and the stressor is documented, numeric water quality criteria (NWQC) for nitrogen or phosphorous could potentially be derived by selecting the stressor value that corresponds to the selected biological threshold (USEPA 2010a). This process facilitates the development of NWQC for nitrogen or phosphorus that are explicitly associated with levels of biological condition supportive of designated aquatic life uses. Developing these relationships at multiple scales including landscape-scale biological assessments will facilitate linking state and tribal water quality standards with both watershed and national estuary programs (Cicchetti and Greening 2011, USEPA In draft).

Biological Assessment at Multiple Scales in Estuaries and Coastal Waters

A large body of estuarine work has been done in index development and in application of habitat-level biological assessments. For example, approaches have been developed for salt marshes, soft-bottom benthic invertebrate communities and seagrass beds (USEPA 2000b). As a supplement to these efforts, several environmental programs such as EPA's national estuary programs (NEPs) are working together with U.S. Environmental Protection Agency (EPA) Office of Research and Development to develop landscape-scale biological assessment tools to evaluate and understand large-scale changes that have occurred to multiple habitats over long time periods and to integrate them into management in conjunction with existing habitat-level biological assessment tools. Specifically, the Tampa Bay, Narragansett Bay, and Mobile Bay Estuary programs are evaluating complementary application of the BCG to estuaries at the individual habitat level of biological assessment and at the landscape level of biological assessment, for managing estuaries and watersheds at the spatial scale of the entire waterbody.

Definitions:

- Habitat-level biological assessments—Evaluations of biological condition that consider biological communities within a defined habitat or suite of habitats (see Frey 1975; Karr and Dudley 1981).
- Landscape-level biological assessments—Evaluations of biological condition that consider and attempt to integrate biological processes, multiple biological habitats, or multiple

biological communities within a defined landscape, waterbody, watershed, or waterbody type. The extent or arrangement (or both) of multiple biological habitats in a defined waterbody type.

- Both of these types of assessments can apply at a wide range of spatial scales, from a single area or subembayment to a larger waterbody, state, region, or nation.

As an example of landscape level assessment, one method in development considers the habitat landscape or biotope mosaic. A *biotope* is an area that is relatively uniform in physical structure and is identified by a dominant biota (Madden et al. 2009; Davies et al. 2004). Biotopes in estuaries include seagrass beds, salt marshes, coral reefs, clam flats, and more. Biotopes are a foundation of many recent habitat classification schemes, including the Coastal and Marine Ecological Classification Standard, which has been sponsored by the Federal Geographic Data Committee, and the European Nature Information System (Davies et al. 2004). Arrangements of biotopes provide species with spawning grounds, nurseries, refuge, sustenance, and other vital needs; such arrangements are particularly critical for larger mobile species and for species that move among biotopes at different stages of their life. The areas and arrangements of biotopes in a waterbody are affected by the full range of anthropogenic stressors, including nitrogen and phosphorus pollution, toxics, shoreline development, and sediment loads. Because biotopes are inherently a biological component, NEPs are developing approaches for biological assessment that consider areas and distributions of biotopes and biotope landscapes at the whole-estuary scale, combining the landscape-level tools with more resolved habitat-level tools. Tampa Bay, Narragansett Bay, and other NEPs are working on these multi-level BCG approaches. Additionally, the Mobile Bay NEP is exploring how to incorporate the concept of ecosystem services in development of a Biological Condition Gradient for the estuary. Current efforts in Tampa Bay and Narragansett Bay are briefly discussed below.

Tampa Bay Estuary Program

The Tampa Bay Estuary Program (TBEP) initiated a system-wide management framework in the 1990s that developed estuarine habitat restoration and protection goals to support estuarine-dependent species and the habitat landscapes they require (e.g., the extent of seagrass beds, mangrove forests, *Spartina* marshes, *Salicornia* marshes, and low-salinity marshes).

Although the term *biotope* was not used, the framework employed the basic concepts of biotope extent and distribution to evaluate condition of the waterbody, comparing current condition to a more naturally occurring condition that existed at a relatively undisturbed point in the past. This information supported the development of environmental protection and restoration goals for the waterbody and watershed that move the estuary closer to those more naturally occurring conditions. This approach was combined with habitat-level work, including water quality modeling to predict seagrass health, benthic macroinvertebrate surveys, and more. Tampa Bay has recovered many hundreds of acres of high-value biotopes (Cloern 2001; Duarte 2009). TBEP is now working with other NEPs to develop those approaches into transferable biological assessment tools using concepts from the BCG. The methods used by TBEP, together with their application to biological assessment at landscape scales, are discussed in Cicchetti and Greening (2011).

Narragansett Bay Estuary Program

The Narragansett Bay Estuary Program (NBEP) and partners, benefiting from the Tampa Bay experience, are developing a suite of biological assessment tools to apply on a range of biological levels and spatial scales. A pilot program in Greenwich Bay, a sub-estuary of Narragansett Bay, has examined macroinvertebrate communities and biotopes in the context of the BCG using historical documentation of early stressor levels and ecosystem conditions to recreate a biological baseline. The project is especially pertinent to highly altered systems where it is often impossible to find undisturbed or minimally disturbed conditions. To characterize the biological responses to increasing stress, the study identified current, recent and historical stressors to Greenwich Bay benthos, including water quality (e.g., hypoxia), sediment metals, nutrients (i.e., nitrogen-loading), and hydrodynamics (including dredging and shoreline modification), terrestrial runoff, storms, and temperature. Changes in these parameters through time were summarized. A critical but challenging aspect of the project was to establish a reference level, or minimally disturbed endpoint. Target reference levels derived from historical baselines can be problematic because (1) they are difficult to calibrate with current ecosystem status, (2) ecosystems were as dynamic in the past as they are today, and (3) climate change and the degree of anthropogenic influence can render these endpoints unattainable. However, Greenwich Bay is fortunate in having available a significant amount of cultural and scientific historical data; although much of the information is qualitative, even qualitative differences in the biological indicators can be useful for defining a minimally disturbed endpoint. Ecological timeline data were overlaid with a detailed cultural timeline in order to associate changes in biological indicators with changes in human activities. Records of significant storms and climate trends gave broader context to ecological observations. The combined cultural and ecological timeline suggest when thresholds in the biological indicators may have been exceeded.

Because nutrient pollution is a major stressor in Narragansett Bay, the tools consider habitats and landscapes that are sensitive to (and diagnostic of) nutrient stress. At the habitat scale, NBEP and EPA's Atlantic Ecology Division (Narragansett) are developing approaches for biological assessment of macroinvertebrates in deeper subtidal areas, camera-based approaches to examine biology in deeper subtidal areas, and approaches for evaluating seagrass and microalgae as tools to better manage nutrient inputs to the waterbody and watershed. The overall project goal is to develop an estuarine framework that can apply at multiple scales and levels using several methods of biological assessment, all brought together with the "common language" of the BCG.

Transferability to Freshwater Aquatic Ecosystems

By performing biological assessments and developing BCGs at multiple spatial scales and levels of biological organization in estuaries and coral reef ecosystems, EPA, NOAA, and their NEP partners will better understand the interactions among biological communities with system-level processes that define and regulate ecosystems, and will be able to assess the cumulative effects of multiple stressors over large spatial scales and over longer periods of time (e.g., decadal). The results of this work are expected to be adapted to large and complex freshwater systems, such as braided river networks, lakes, and large rivers and their attendant watersheds. In river systems, for example, EPA's Ecological Exposure Research Division (Cincinnati) is developing geographic information system- (GIS-) based tools to classify and characterize natural variability in watersheds and concurrently developing watershed-scale models integrating habitat and landscape biological assessments of classified river systems, incorporating main channel and lateral slackwaters (bays, side channels, and backwaters) with the floodscape (isolated oxbows, lakes, wetlands, and usually dry alluvial floodplains). A major component of this work focuses on defining critical ecological thresholds, or tipping points, of ecological condition and function in river systems in response to multiple stressors in watersheds at multiple spatial and temporal scales.

Tools such as these could support watershed and basin wide management and planning, enabling state, tribal and local resource managers to: 1) account for more of the natural variability within and across river systems, watershed and regions; 2) relate changes in stressors exposure to changes in biological (and functional) condition at both a watershed and system-wide level; and, 3) facilitate the extrapolation of findings from one system and/or watershed to other similarly located or functioning systems.

3.17. Partnerships in the Protection of Oregon's Coho Salmon

Abstract

Assessment of biological conditions in Oregon's Coast Coho Evolutionarily Significant Unit (ESU) has provided state agencies with valuable information that can be used to improve protection of coho salmon. Oregon Department of Environmental Quality and Oregon Department of Fish and Wildlife are using monitoring data to examine several indicators—temperature and fine sediments—that have been identified as potential causes of coho population decline in the state. Findings show that the two monitoring areas with the highest biological condition also showed the lowest evidence of stress from temperature and fine sediment. National Oceanic and Atmospheric Administration's Fisheries Division has also been able to use biological information to support a decision to list coho as *threatened* and to designate the Oregon Coast Coho ESU as a critical habitat.

Introduction

For more than a decade, state and federal agencies have been working to halt the decline of coho salmon in Oregon. In 1997 Oregon implemented the Oregon Plan for Salmon and Watersheds, a step toward reversing the decline of coho salmon in Oregon coastal streams. In response, Oregon Department of Environmental Quality (ODEQ) and Oregon Department of Fish and Wildlife (ODFW) began expanded monitoring in Oregon coastal streams to gather information on the status of water quality and watershed health indicators identified as potential causes for declining populations of Oregon coastal coho salmon (State of Oregon 1997).

In 2005 ODEQ and ODFW assessed the information collected on the factors for the decline of coho and evaluated the relative importance of each factor to the continued viability of Oregon's coastal coho runs into the future.

Specifically, ODEQ and ODFW assessed data for the Oregon Coast Coho Evolutionarily Significant Unit (ESU). The Oregon Coast Coho ESU is in western Oregon, spanning approximately three-quarters of the coastline with the Pacific Ocean and contains more than 9,000 miles of rivers and streams. Most of the stream miles (more than 80 percent) are small, wadeable streams (1st through 3_{rd} order). Two hundred and eighty-three randomly selected sites were characterized throughout the ESU, ranging from 61 to 86 sites per monitoring area. Specifically, data were analyzed for four monitoring areas nested within the ESU (North Coast, Mid-Coast, Mid-South Coast, and Umpqua).

In 2007 ODFW released the final draft of the *Oregon Coast Coho Conservation Plan* (State of Oregon 2007), which outlines Oregon's strategy to ensure the continued viability of threatened coastal coho salmon runs. Part of the plan identifies the need for higher-resolution monitoring of water quality and macroinvertebrates in the Oregon Coast Coho ESU (Lawson et al. 2007). Because of the ability of macroinvertebrates to integrate the effects of water quality and habitat stressors—and limited resources for comprehensive monitoring—ODEQ and ODFW agreed that macroinvertebrates would be used to relate water quality and overall watershed condition in the ESU. In 2008, National Oceanic and Atmospheric Administration (NOAA) Fisheries Division used the information in its final decision to re-list Oregon coastal coho as *threatened* under the Endangered Species Act.

Assessment of Biological Condition

In 2006–2007 ODEQ and ODFW jointly collected and analyzed macroinvertebrate data in the ESU. They evaluated biological condition for each of four monitoring areas in the ESU. Macroinvertebrates were also used as a screening tool to determine the relative contributions of temperature and fine sediment as stressors to biological condition.

A multivariate predictive model, PREDATOR, was used to assess the biological condition of wadeable streams throughout Oregon (Hubler 2008). The model compares observed taxa with expected taxa to generate an observed/expected (O/E) taxa ratio. Scores of less than 1.0 have fewer taxa at a site than were predicted by the model, representing a loss of native reference taxa richness. Benchmarks based on the distribution of O/E scores at reference sites were used to classify the samples into one of the three following biological condition classes: least disturbed, moderately disturbed, and most disturbed (Table 3-12).

Table 3-12. Biological benchmarks

Biological condition class	O/E	Taxa loss
Least disturbed	> 0.91	8% or less
Moderately disturbed	0.86–0.91	9%–14%
Most disturbed	< 0.86	15% or more

Subsequent monitoring showed that approximately 50 percent of the streams could be classified as least disturbed (equivalent to reference), while almost 40 percent of streams in the ESU had macroinvertebrates in most disturbed conditions (missing a considerable amount of reference taxa). The four monitoring units showed different relative proportions of condition classes. The Mid-Coast monitoring area had the largest proportion of sites in highest biological condition with 69 percent of sites in least disturbed condition and 17 percent of sites in most disturbed condition. The Umpqua monitoring unit showed only about one-quarter of sites in least disturbed conditions and approximately two-thirds of sites in most disturbed conditions. That information, along with stressor information for each monitoring unit, became very important in developing the stressor-response model. The information was used to try to identify the relative importance of two key (NPS) stressors to macroinvertebrate conditions in the Oregon Coastal Coho ESU.

Stressor-Response Model

The relationships among macroinvertebrate abundances and environmental variables (seasonal maximum temperature and percent fines) were used to model the optimum conditions for each taxon. These optimal conditions were then used to infer the overall assemblage preference for temperature and fine sediments of any site using a macroinvertebrate sample alone (Huff et al. 2006). Benchmarks were established to identify sites where temperature or fine sediments or both can be at levels considered to be stressful to the macroinvertebrate assemblages. Temperature stress (TS) values above 18 °C were considered temperature stressed, as it relates directly to the WQS set to protect salmon and trout rearing and migration. Fine sediment stress (FSS) values above 10 percent were considered sediment stressed because that value has been shown to negatively affect macroinvertebrates in mountain streams (Bryce et al. 2010).

The North Coast monitoring area showed the lowest levels of TS (36 percent of sites) and FSS (22 percent). The Mid-Coast monitoring area showed approximately half of the sites as stressed for both temperature and fine sediment, despite showing the highest percentage of sites in least disturbed biological condition. Both the Mid-South and Umpqua monitoring areas showed two-thirds or more of the sites to be stressed for both temperature and fine sediment. Apart from the North Coast, stresses to the macroinvertebrate assemblages from temperature and fine sediments appear to be equivalent.

Conclusion

Biological data and stressor-response relationships were used as the basis for several findings. First, NOAA was able to make a decision to list coho as *threatened* and to designate the Oregon Coast ESU as a critical habitat. Second, several general trends were observed in the assessment of the macroinvertebrate data collected and assessed. The two monitoring areas with the highest biological condition (North Coast and Mid-Coast) showed the lowest evidence of stress from temperature and fine sediment. The Mid-South Coast and Umpqua monitoring areas showed higher levels of stress and lower biological condition (substantially so in the Umpqua). That information can be used in developing management plans for ESU monitoring areas or basins. Much emphasis has been placed on improving the temperature conditions in Oregon's streams and rivers, while less work has gone into developing sediment management plans. The data presented here suggest that excess fine sediments are affecting biological conditions in the ESU on a scale similar to that of temperature.

Finally, the monitoring project is an example of two state agencies working together to implement a monitoring program that is cost-effective by addressing both agencies' needs for information. For ODFW, the random macroinvertebrate, water quality, and habitat sampling protocol provides critical information on water quality and habitat conditions, which have been identified as limiting factors to coho salmon viability. For ODEQ, the macroinvertebrate sampling in conjunction with the water quality and habitat monitoring provides valuable information on attainment of the designated aquatic life uses for streams.

REFERENCES

ADEQ (Arizona Department of Environmental Quality). 2007. *Technical Support Documentation for the Narrative Biological Standard.* Arizona Department of Environmental Quality, Phoenix, AZ.

ADEQ (Arizona Department of Environmental Quality). 2008. *Biocriteria Implementation Procedures, Draft*. Arizona Department of Environmental Quality, Phoenix, AZ. <http://www.azdeq.gov/environ/water bio.pdf>. Accessed September 2011.

ADEQ (Arizona Department of Environmental Quality). 2010. *Standard Operating Procedures for Water Quality Monitoring*. Arizona Department of Environmental Quality, Phoenix, AZ. <http://www.azdeq.gov/environ /water>.
Accessed Accessed September 2011.

Aguilar-Perera, A. and R.S. Appeldoorn. 2007. Variation in juvenile fish density along mangrove-seagrasscoral reef continuum in SW Puerto Rico. *Marine Ecology Progress Series* 348:139–148.

Bryce, S.A., G.A. Lomnicky, and P.R. Kaufmann. 2010. Protecting sediment-sensitive aquatic species in mountain streams through the application of biologically based streambed sediment criteria. *Journal of the North American Benthological Society* 29(2):657–672.

BLM (Bureau of Land Management). 2010. *John Day River*. Department of the Interior, Bureau of Land Management. <http://www.blm.gov/or /resources>. Accessed September 2011.

Bradley, P., L. Fore, W. Fisher, and W. Davis. 2010. *Coral Reef Biological Criteria: Using the Clean Water Act to Protect a National Treasure*. EPA-600-R-10-054. U.S. Environmental Protection Agency, Office of Research and Development, Narragansett, RI. <http://cfpub.epa.gov/si/sipublic recordreport.cfm?dirEntryId=223392>. Accessed September 2011.

Carlisle, D.M., D.M. Wolock, and M.R. Meador. 2010. Alteration of streamflow magnitudes and potential ecological consequences: A multiregional assessment. *Frontiers in Ecology and the Environment* 2010:doi.1890/100053. <http://water FLowAlterationUS.pdf>. Accessed August 2011.

Christensen, J.D, C.F.G. Jeffery, C. Caldow, M.E. Monaco, M.S. Kendall, and R.S. Appeldoorn. 2003. Cross-shelf habitat utilization patterns of reef fishes in Southwestern Puerto Rico. *Gulf and Caribbean Research* 14(2):9–27.

Cicchetti, G., and H. Greening. 2011. Estuarine biotope mosaics and habitat management goals: An application in Tampa Bay, Florida, USA. *Estuaries and Coasts* (May):1–15. doi: 10.1007/s12237- 011-9408-4.

Cloern, J.E. 2001. Our evolving conceptual model of the coastal eutrophication problem. *Marine Ecology Progress Series* 210:223–253.

Davies, C.E., D. Moss, and M.O. Hill. 2004. *EUNIS Habitat Classification Revised 2004.* Report to European Environment Agency, European Topic Center on Nature Protection and Biodiversity.

Davies, S.P., and S.K. Jackson. 2006. The biological condition gradient: A descriptive model for interpreting change in aquatic ecosystems. *Ecological Applications* 16(4):1251–1266.

Duarte, C.M. 2009. Coastal eutrophication research: A new awareness. *Hydrobiologia* 629:263–269.

Fisher, W.S. 2007. *Stony Coral Rapid Bioassessment Protocol.* EPA/600/R-06/167. U.S. Environmental Protection Agency. <http://www.epa.gov/bioiweb1/pdf/EPA-600-R-06-167StonyCoralRBP.pdf>. Accessed August 2011.

Fisher, W.S., W.P. Davis, R.L. Quarles, J. Patrick, J.G. Campbell, P.S. Harris, B.L. Hemmer, and M. Parsons. 2007. Characterizing coral condition using estimates of three-dimensional coral surface area. *Environmental Monitoring and Assessment* 125:347–360.

Fisher W.S., L.S. Fore, A. Hutchins, R.L. Quarle, J.G. Campbell, C. LoBue, and W. Davis. 2008. Evaluation of stony coral indicators for coral reef management. *Marine Pollution Bulletin* 56:1737–1745.

Frey, D. 1975. Biological integrity of water: an historical approach. In *The Integrity of Water,* ed. R.K.

Ballentine, and L.J. Guarraia, pp. 127–140. Proceedings of a Symposium, March 10–12, 1975. U.S. Environmental Protection Agency, Washington, DC.

H. John Heinz III Center for Science, Economics, and the Environment. 2008. *The State of the Nation's Ecosystems.* Island Press, Washington, DC. <http://www.heinzctr.org/ecosystems>. Accessed June 2011.

Hubler, S. 2008. PREDATOR: *Development and Use of RIVPACS-type Macroinvertebrate Models to Assess the Biotic Condition of Wadeable Oregon Streams.* DEQ08-LAB-0048-TR. Oregon Department of Environmental Quality, Hillsboro, OR.

Huff D.D., S. Hubler, V. Pan, and D. Drake. 2006. *Detecting Shifts in Macroinvertebrate Community Requirements: Implicating Causes of Impairment in Streams.* DEQ06-LAB-0068-TR. Oregon Department of Environmental Quality, Hillsboro, OR.

IDNR (Iowa Department of Natural Resources). No date. *North Fork Maquoketa River* (factsheet). Iowa

Department of Natural Resources. Available upon request from Iowa Department of Natural Resources, 502 East 9th Street Des Moines, IA, 50319-0034 or Jeff.Berckes@dnr.iowa.gov.

IDNR (Iowa Department of Natural Resources). 2006. *Stressor Identification: North Fork Maquoketa River, Iowa.* Iowa Department of Natural Resources. <http://www.iowadnr.gov/Environment/WaterQuality/WatershedImprovement/WatershedResearchData/WaterImprovement Plans/PublicMeetingsPlans.aspx>. Document listed as nforkmaqsi.pdf under Final Water Quality Improvement Plans. Accessed September 2011.

IDNR (Iowa Department of Natural Resources). 2007. *Total Maximum Daily Loads for Sediment, Nutrients, and Ammonia: North Fork Maquoketa River, Dubuque County, Iowa.* Iowa Department of Natural Resources. <http://www.iowadnr.gov/Environment/WaterQuality/WatershedImprovement/WatershedResearchData/WaterImprovementPlans/PublicMeetings Plans.aspx>. Document listed as northforkmaq.pdf under Final Water Quality Improvement Plans. Accessed September 2011.

Jelks, H.L., S.J. Walsh, N.M. Burkhead, S. Contreras-Balderas, E. Diaz-Pardo, D.A. Hendrickson, J. Lyons, N.E. Mandrak, F. McCormick, J.S. Nelson, S.P. Plantania, B.A. Porter, C.B. Renaud, J.J. SchmitterSoto, E.B. Taylor, and M.L. Warren, Jr. 2008. Conservation status of imperiled North American freshwater and diadromous fishes. *Fisheries* 33(8):372–407.

Karr, J.R., and D.R. Dudley. 1981. Ecological perspectives on water quality goals. *Environmental Management* 5:55–68.

Lawson, P.W., E.P. Bjorkstedt, M.W. Chilcote, C.W. Huntington, J.S. Mills, K.M.S. Moore, T.E. Nickelson, G.H. Reeves, H.A. Stout, T.C. Wainwright, and L.A. Weitkamp. 2007. *Identification of historical populations of Coho salmon (*Oncorhynchus kisutch*) in the Oregon coast evolutionarily significant unit.* U.S. Department of Commerce, NOAA Tech. Memo. NMFS-NWFSC-79, 129 p. <http://www.nwfsc.noaa.gov /assets 08302007104459HistPopsCohoTM79Final.pdf>. Accessed September 2011.

Madden, C.J., K. Goodin, R.J. Allee, G. Cicchetti, C. Moses, M. Finkbeiner, and D. Bamford. 2009. *Coastal and Marine Ecological Classification Standard.* National Oceanic and Atmospheric Administration and NatureServe.

McField, M., and P.R. Kramer. 2007. *Healthy Reefs for Healthy People: A Guide to Indicators of Reef Health and Social Well-being in the Mesoamerican Reef Region.* With contributions by M. Gorrez and M. McPherson.

Meynecke, J.O., S.Y. Lee, and N.C. Duke. 2008. Linking spatial metrics and fish catch reveals the importance of coastal wetland connectivity to inshore fisheries in Queensland, Australia. *Biological Conservation* 141:981–996.

Mumby, P.J., A.J. Edwards, J.E. Arias-Gonzalez, K.C. Lindeman, P.G. Blackwell, A. Gall, M.I. Gorczynska, A.R. Harborne, C.L. Pescod, H. Renken, C.C.C. Wabnitz, and G. Llewellyn. 2004. Mangroves enhance the biomass of coral reef fish communities in the Caribbean. *Nature* 427(6974):533–536.

Mumby, P.J., K. Broad, D.R. Brumbaugh, C.P. Dahlgren, A.R. Harborne, A. Hastings, K.E. Holmes, C.V. Kappel, F. Micheli, and J.N. Sanchirico. 2008. Coral reef habitats as surrogates of species, ecological functions, and ecosystem services. *Conservation Biology* 22: 941–951.

Ohio EPA (Ohio Environmental Protection Agency). 1996. *Justification and Rationale for Revisions to the Dissolved Oxygen Criteria in the Ohio Water Quality Standards*. OEPA Technical Bulletin MAS/1995-12-5, State of Ohio Environmental Protection Agency, Division of Surface Water, Columbus, OH.

Omernik, K.M. 1987. Ecoregions of the conterminous United States. *Annals of the Association of American Geographers* 77:118–125.

PA DEP (Pennsylvania Department of Environmental Protection). 2009. *A Benthic Index of Biotic Integrity for Wadeable Freestone Riffle-Run Streams in Pennsylvania*. Pennsylvania Department of Environmental Protection. <http://files.dep.state WaterQualityPortalFiles/ibi rifflerun2009.pdf>. Accessed September 2011.

Plafkin, J.L., M.T. Barbour, K.D. Porter, S.K. Gross, and R.M. Hughes. 1989. *Rapid Bioassessment Protocols for Use in Streams and Rivers: Benthic Macroinvertebrates and Fish*. EPA-444-4-89-001. U.S. Environmental Protection Agency, Washington, DC.

Poff, N.L., and J.K.H. Zimmerman. 2010. Ecological responses to altered flow regimes: A literature review to inform the science and management of environmental flows. *Freshwater Biology* 55:194–205.

Rankin, E.T. 1989. *The Qualitative Habitat Evaluation Index (QHEI), Rationale, Methods, and Application*. Ohio Environmental Protection Agency, Division of Water Quality Planning and Assessment, Ecological Assessment Section, Columbus, OH.

Rankin, E.T. 1995. The use of habitat indices in water resource quality assessments. In *Biological Assessment and Criteria: Tools for Water*

Resource Planning and Decision Making, ed. W.S. Davis, and P. Simon, pp. 181–208. Lewis Publishers, Boca Raton, FL.

Sale, P.F., P. Jacob, and J.P. Kritzer. 2008. Connectivity: What it is, how it is measured, and why it is important for management of reef fishes. In *Caribbean Connectivity: Implications for Marine Protected Area Management*, ed. R. Grober-Dunsmore and B.D. Keller, pp. 16–30. Proceedings of a Special Symposium, 9–11 November 2006, 59th Annual Meeting of the Gulf and Caribbean Fisheries Institute, Belize City, Belize. Marine Sanctuaries Conservation Series ONMS-08-07. U.S. Department of Commerce, National Oceanic and Atmospheric Administration, Office of National Marine Sanctuaries, Silver Spring, MD.

Spindler, P.H. 2001. *Macroinvertebrate Community Distribution among Reference Sites in Arizona.* OFR 00-05. Arizona Department of Environmental Quality, Phoenix, AZ.

State of Oregon. 1997. *The Oregon Plan: Oregon Coastal Salmon Restoration Initiative.* <http://ir.library.oregonstate.edu/dspace/handle/1957/4983>. Accessed September 2011.

State of Oregon. 2007. *Oregon Coast Coho Conservation Plan for the State of Oregon*. Prepared by the Oregon Department of Fish and Wildlife. <http://www.wrd.state>. Accessed September 2011.

Stoddard, J.L., D.P. Larsen, C.P. Hawkins, R.K. Johnson, and R.H. Norris. 2006. Setting Expectations for the Ecological Condition of Streams: The Concept of Reference Condition. *Ecological Applications* 16(4):1267–1276.

USACE (U.S. Army Corps of Engineers). 2002. Memo: Grant of Section 401 Certification Authorization of discharges of dredged or fill material to various waters of the state [Ohio] for Nationwide Permits as detailed in the January 15, 2002, *Federal Register* (Volume 67, Number 10). U.S. Army Corps of Engineers, CECW-OR, Washington, DC.

USEPA (U.S. Environmental Protection Agency). 1990. *Biological Criteria: National Program Guidance for Surface Waters.* EPA-440-5-90-004. U.S. Environmental Protection Agency, Washington, DC. <http://www.epa.gov /bioindicators>. Accessed September 2011.

USEPA (U.S. Environmental Protection Agency). 1991a. *Technical Support Document for Water Quality-based Toxics Control.* EPA -5052-90-001. U.S. Environmental Protection Agency, Washington, DC. <http://www.epa.gov/npdes/pubs/owm0264.pdf>. Accessed September 2011.

USEPA (U.S. Environmental Protection Agency). 1991b. *Policy on the Use of Biological Assessments and Criteria in the Water Quality Program.* U.S. Environmental Protection Agency, Washington, DC. <http://www.epa.gov /bioiweb1/pdf/PolicyonBiologicalAssessmentsandCriteria.pdf>. Accessed September 2011.

USEPA (U.S. Environmental Protection Agency). 1994. *Water Quality Standards Handbook.* 2nd ed. EPA-823-B-94-005. U.S. Environmental Protection Agency, Washington, DC. <http://water>. Accessed September 2011.

USEPA (U.S. Environmental Protection Agency). 1999. *Rapid Bioassessment Protocols for Use in Streams and Wadeable Rivers: Periphyton, Benthic Macroinvertebrates and Fish.* 2nd ed. EPA 841-B-99- 002. U.S. Environmental Protection Agency, Office of Water, Washington, DC. <http://water>. Accessed September 2011.

USEPA (U.S. Environmental Protection Agency). 2000a. *Stressor Identification Guidance Document.* EPA-822-B-00-025. U.S. Environmental Protection Agency, Washington, DC. <http://www.epa.gov /waterscience/biocriteria/stressors>. Accessed September 2011.

USEPA (U.S. Environmental Protection Agency). 2000b. *Estuarine and Coastal Marine Waters: Bioassessment and Biocriteria Technical Guidance*. EPA 822-B-00-024. U.S. Environmental Protection Agency, Office of Water, Washington, DC. <http://water upload/20090422 biocriteriaStatesestuariesestuaries-2.pdf>. Accessed September 2011.

USEPA (U.S. Environmental Protection Agency). 2002. *Summary of Biological Assessment Programs and Biological Criteria Development for States, Tribes, Territories and Interstate Commissions: Streams and Wadeable Rivers*. EPA-822-R-02-048. U.S. Environmental Protection Agency, Office of Water, Washington, DC. <http://www.epa.gov /bioiweb1/html/programsummary.html>. Accessed September 2011.

USEPA (U.S. Environmental Protection Agency). 2003. *Elements of a State Water Monitoring and Assessment Program*. EPA 841-B-03-003. U.S. Environmental Protection Agency, Washington, DC. <http://water>. Accessed September 2011.

USEPA (U.S. Environmental Protection Agency). 2006. *Wadeable Streams Assessment*. EPA 841-B-06- 002. U.S. Environmental Protection Agency, Office of Research and Development and Office of Water, Washington, DC. <http://water>. Accessed September 2011.

USEPA (U.S. Environmental Protection Agency). 2007. *Watershed-based National Pollutant Discharge Elimination System (NPDES) Permitting*

Technical Guidance. EPA 833-B-07-004. U.S. Environmental Protection Agency, Office of Wastewater Management, Washington, DC. <http://www.epa.gov/npdes/pubs/watershed techguidance.pdf>. Accessed September 2011.

USEPA (U.S. Environmental Protection Agency). 2010a. *Using Stressor-response Relationships to Derive Numeric Nutrient Criteria*. EPA-820-2-10-001. U.S. Environmental Protection Agency. <http://water .pdf>. Accessed September 2011.

USEPA (U.S. Environmental Protection Agency). 2010b. *Causal Analysis/Diagnosis Decision Information System (CADDIS)*. Office of Research and Development, Washington, DC. <http://www.epa.gov /caddis>. Website last updated September 23, 2010. Accessed September 2011.

USEPA (U.S. Environmental Protection Agency). 2010c. *Implications of Climate Change for State Bioasssessment Programs and Approaches to Account for Effects (External Review Draft)*. U.S. Environmental Protection Agency, Office of Research and Development, Washington, DC. <http://cfpub.epa.gov/ncea/global/recordisplay.cfm?deid=233810>. Accessed September 2011.

USEPA (U.S. Environmental Protection Agency). 2012. *Recovery Potential Screening*. U.S. Environmental Protection Agency, Office of Wetlands, Oceans and Watersheds. Washington, DC. <http://www.epa.gov /recovery> Accessed January 2012.

USEPA (U.S. Environmental Protection Agency). In draft. *Identifying and Protecting Healthy Watersheds: A Technical Guide* (Draft). U.S. Environmental Protection Agency, Office of Wetlands, Oceans, and Watersheds. Washington, DC. <http://water>. In process of finalization. Release expected 2012.

Walter, R.C., and D.J. Merritts. 2008. Natural streams and the legacy of water-powered mills. *Science* 319:299–304.

Yoder C.O., and J.E. DeShon. 2003. Using biological response signatures in a framework of multiple indicators to assess and diagnose causes and sources of impairments to aquatic assemblages in selected Ohio rivers and streams. In *Biological Response Signatures: Indicator Patterns using Aquatic Communities*, ed. T.P. Simon, pp. 23–81. CRC Press, Boca Raton, FL.

Yoder, C.O., and E.T. Rankin. 1995. Biological Response Signatures and the Area of Degradation Value: New Tools for Interpreting Multimetric Data. In *Biological Assessment and Criteria: Tools for Water Resource*

Planning and Decision Making, ed. W.S. Davis and P. Simon, pp. 263–286. Lewis Publishers, Boca Raton, FL.

GLOSSARY

aquatic assemblage: An association of interacting populations of organisms in a given waterbody; for example, fish assemblage or a benthic macroinvertebrate assemblage.

aquatic community: An association of interacting assemblages in a waterbody, the biotic component of an ecosystem.

aquatic life use: A beneficial use designation in which the waterbody provides, for example, suitable habitat for survival and reproduction of desirable fish, shellfish, and other aquatic organisms.

attribute: The measurable part or process of a biological system.

benthic macroinvertebrates or benthos: Animals without backbones, living in or on the sediments, of a size large enough to be seen by the unaided eye and which can be retained by a U.S. Standard no. 30 sieve (28 meshes per inch, 0.595-mm openings); also referred to as benthos, infauna, or macrobenthos.

best management practice: An engineered structure or management activity, or combination of those, that eliminates or reduces an adverse environmental effect of a pollutant.

biological assessment or bioassessment: An evaluation of the biological condition of a waterbody using surveys of the structure and function of a community of resident biota.

biological criteria or biocriteria: Narrative expressions or numeric values of the biological characteristics of aquatic communities based on appropriate reference conditions; as such, biological criteria serve as an index of aquatic community health.

biological indicator or bioindicator: An organism, species, assemblage, or community characteristic of a particular habitat, or indicative of a particular set of environmental conditions.

biological integrity: The ability of an aquatic ecosystem to support and maintain a balanced, adaptive community of organisms having a species composition, diversity, and functional organization comparable to that of natural habitats in a region.

biological monitoring or biomonitoring: Use of a biological entity as a detector and its response as a measure to determine environmental conditions; ambient biological surveys and toxicity tests are common biological monitoring methods.

biological survey or biosurvey: Collecting, processing, and analyzing a representative portion of the resident aquatic community to determine its structural and/or functional characteristics.

biotope: An area that is relatively uniform in physical structure and that is identified by a dominant biota.

clean Water Act: The act passed by the U.S. Congress to control water pollution (formally referred to as the Federal Water Pollution Control Act of 1972). Public Law 92-500, as amended. 33 U.S.C. 1251 *et seq.*

clean Water Act 303(d): This section of the act requires states, territories, and authorized tribes to develop lists of impaired waters for which applicable WQS are not being met, even after point sources of pollution have installed the minimum required levels of pollution control technology. The law requires that the jurisdictions establish priority rankings for waters on the lists and develop TMDLs for the waters. States, territories, and authorized tribes are to submit their lists of waters on April 1 in every even-numbered year.

Clean Water Act 305(b): Biennial reporting requires description of the quality of the nation's surface waters, evaluation of progress made in maintaining and restoring water quality, and description of the extent of remaining problems.

criteria: Elements of state water quality standards, expressed as constituent concentrations, levels, or narrative statements, representing a quality of water that supports a particular use. When criteria are met, water quality will generally protect the designated use.

designated uses: Those uses specified in WQS for each waterbody or segment whether or not they are being attained.

disturbance: Human activity that alters the natural state and can occur at or across many spatial and temporal scales.

ecological integrity: The condition of an unimpaired ecosystem as measured by combined chemical, physical (including physical habitat), and biological attributes. Ecosystems have integrity when they have their native components (plants, animals and other organisms) and processes (such as growth and reproduction) intact.

ecoregion: A relatively homogeneous ecological area defined by similarity of climate, landform, soil, potential natural vegetation, hydrology, or other ecologically relevant variables.

function: Processes required for normal performance of a biological system (may be applied to any level of biological organization).

guild: A group of organisms that exhibit similar habitat requirements and that respond in a similar way to changes in their environment.

historical data: Data sets from previous studies, which can range from handwritten field notes to published journal articles.

index of biological/biotic integrity: An integrative expression of site condition across multiple metrics; an IBI is often composed of at least seven metrics.

invasive species: A species whose presence in the environment causes economic or environmental harm or harm to human health. Native species or nonnative species can show invasive traits, although that is rare for native species and relatively common for nonnative species. (Note that this term is not included in the biological condition gradient [BCG].)

least disturbed condition: The best available existing conditions with regard to physical, chemical, and biological characteristics or attributes of a waterbody within a class or region. Such waters have the least amount of human disturbance in comparison to others in the waterbody class, region, or basin. Least disturbed conditions can be readily found but can depart significantly from natural, undisturbed conditions or minimally disturbed conditions. Least disturbed condition can change significantly over time as human disturbances change.

maintenance of populations: Sustained population persistence; associated with locally successful reproduction and growth.

metric: A calculated term or enumeration that represents some aspect of biological assemblage, function, or other measurable aspect and is a characteristic of the biota that changes in some predictable way with increased human influence.

minimally disturbed condition: The physical, chemical, and biological conditions of a waterbody with very limited, or minimal, human disturbance.

multimetric index: An index that combines indicators, or metrics, into a single index value. Each metric is tested and calibrated to a scale and transformed into a unitless score before being aggregated into a multimetric index. Both the index and metrics are useful in assessing and diagnosing ecological condition. See index of biological/biotic integrity (IBI).

narrative biological criteria: Written statements describing the structure and function of aquatic communities in a waterbody that support a designated aquatic life use.

native: An original or indigenous inhabitant of a region; naturally present.

nonnative or intentionally introduced species: With respect to an ecosystem, any species that is not found in that ecosystem; species introduced or spread from one region of the United States to another outside their normal range are nonnative or non-indigenous, as are species introduced from other continents.

numeric biological criteria: Specific quantitative measures of the structure and function of aquatic communities in a waterbody necessary to protect a designated aquatic life use.

periphyton: A broad organismal assemblage composed of attached algae, bacteria, their secretions, associated detritus, and various species of microinvertebrates.

rapid bioassessment protocols: Cost-effective techniques used to survey and evaluate the aquatic community to detect aquatic life impairments and their relative severity.

reference condition (biological integrity): The condition that approximates natural, unaffected conditions (biological, chemical, physical, and such) for a waterbody. Reference condition (biological integrity) is best determined by collecting measurements at a number of sites in a similar waterbody class or region undisturbed by human activity, if they exist. Because undisturbed conditions can be difficult or impossible to find, minimally or least disturbed conditions, combined with historical information, models, or other methods can be used to approximate reference condition as long as the departure from natural or ideal is understood. Reference condition is used as a benchmark to determine how much other waterbodies depart from this condition because of human disturbance. See definitions for minimally and least disturbed condition.

reference site: A site selected for comparison with sites being assessed. The type of site selected and the types of comparative measures used will vary with the purpose of the comparisons. For the purposes of assessing the ecological condition of sites, a reference site is a specific locality on a waterbody that is undisturbed or minimally disturbed and is representative of the expected ecological integrity of other localities on the same waterbody or nearby waterbodies.

refugia: Accessible microhabitats or regions in a stream reach or watershed where adequate conditions for organism survival are maintained during circumstances that threaten survival; for example, drought, flood, temperature extremes, increased chemical stressors, habitat disturbance.

sensitive taxa: Taxa intolerant to a given anthropogenic stress; first species affected by the specific stressor to which they are *sensitive* and the last to recover following restoration.

sensitive or regionally endemic taxa: Taxa with restricted, geographically isolated distribution patterns (occurring only in a locale as opposed to a region), often because of unique life history requirements. Can be long-lived, late-maturing, low-fecundity, limited-mobility, or require mutualist relation with other species. Can be among listed endangered/threatened or special concern species. Predictability of occurrence often low; therefore, requires documented observation. Recorded occurrence can be highly dependent on sample methods, site selection, and level of effort.

sensitive - rare taxa: Taxa that naturally occur in low numbers relative to total population density but can make up large relative proportion of richness. Can be ubiquitous in occurrence or can be restricted to certain micro-habitats, but because of low density, recorded occurrence is dependent on sample effort. Often stenothermic (having a narrow range of thermal tolerance) or coldwater obligates; commonly k-strategists (populations maintained at a fairly constant level; slower development; longer life span). Can have specialized food resource needs or feeding strategies. Generally intolerant to significant alteration of the physical or chemical environment; are often the first taxa observed to be lost from a community.

sensitive - ubiquitous taxa: Taxa ordinarily common and abundant in natural communities when conventional sample methods are used. Often having a broader range of thermal tolerance than sensitive or rare taxa. These are taxa that constitute a substantial portion of natural communities and that often exhibit negative response (loss of population, richness) at mild pollution loads or habitat alteration.

stressors: Physical, chemical, and biological factors that adversely affect aquatic organisms.

structure: Taxonomic and quantitative attributes of an assemblage or community, including species richness and relative abundance structurally and functionally redundant attributes of the system and characteristics, qualities, or processes that are represented or performed by more than one entity in a biological system.

taxa: A grouping of organisms given a formal taxonomic name such as species, genus, family, and the like.

taxa of intermediate tolerance: Taxa that compose a substantial portion of natural communities; can be r-strategists (early colonizers with rapid turnover times; boom/bust population characteristics). Can be eurythermal (having a broad thermal tolerance range). Can have generalist or facultative feeding strategies enabling utilization of relatively more diversified food types. Readily collected with conventional sample methods. Can increase in number in waters with moderately increased organic resources and reduced competition but are intolerant of excessive pollution loads or habitat alteration.

tolerant taxa: Taxa that compose a small proportion of natural communities. They are often tolerant of a broader range of environmental conditions and are thus resistant to a variety of pollution- or habitat-induced stresses. They can increase in number (sometimes greatly) in the absence of competition. Commonly r-strategists (early colonizers with rapid turnover times; boom/bust population characteristics), able to capitalize when stress conditions occur; last survivors.

total maximum daily load: The sum of the allowable loads of a single pollutant from all contributing point and nonpoint sources; the calculated maximum amount of a pollutant a waterbody can receive and still meet WQS and an allocation of that amount to the pollutant's source.

toxicity identification evaluation: A set of procedures to identify the specific chemicals responsible for effluent toxicity.

toxicity reduction evaluation: A site-specific study conducted in a stepwise process designed to identify the causative agents of effluent toxicity, isolate the sources of toxicity, evaluate the effectiveness of toxicity control options, and then confirm the reduction in effluent toxicity.

water quality management: Decisions on management activities relevant to a water (nonregulatory) resource, such as problem identification, need for and placement of best management practices, pollution abatement actions, and effectiveness of program activity.

water quality standard: A law or regulation that consists of the designated use or uses of a waterbody, the narrative or numerical water quality criteria (including biological criteria) that are necessary to protect the use or uses of that waterbody, and an antidegradation policy.

whole effluent toxicity: The aggregate toxic effect of an aqueous sample (e.g., whole effluent wastewater discharge) as measured by an organism's response after exposure to the sample (e.g., lethality, impaired growth or reproduction); WET tests replicate the total effect and actual environmental exposure of aquatic life to toxic pollutants in an effluent without requiring the identification of the specific pollutants.

ABBREVIATIONS AND ACRONYMS

ADEQ	Arizona Department of Environmental Quality
BCG	biological condition gradient
BMIBI	benthic macroinvertebrate index of biotic integrity
BMP	best management practice
CADDIS	Causal Analysis/Diagnosis Decision Information System
CT DEP	Connecticut Department of Environmental Protection
CWA	Clean Water Act
CWH	coldwater habitat
EPA	U.S. Environmental Protection Agency
EPT	ephemeroptera, plecoptera, trichoptera taxa
ESU	evolutionarily significant unit
EV	exceptional value (Pennsylvania)
EWH	exceptional warmwater habitat
FIBI	fish index of biotic integrity
FQI	Floristic Quality Index
FSS	fine sediment stress
GIS	geographic information system
GPS	global positioning system
HQ	high-quality (Pennsylvania)
HUC	hydrologic unit code
IBI	index of biological/biotic integrity
IC	impervious cover
ICI	invertebrate community index
IDNR	Iowa Department of Natural Resources
INSTAR	Interactive Stream Assessment Resource
IRG	Integrated Reporting Guidance
LRW	limited resource water
LWH	limited warmwater habitat
MDE	Maryland Department of the Environment
MDEQ	Michigan Department of Environmental Quality
MDNR	Maryland Department of Natural Resources

ME DEP	Maine Department of Environmental Protection
mIBI	modified index of biological integrity
MIwb	modified index of well-being
MPCA	Minnesota Pollution Control Agency
MWH	modified warmwater habitat
NARS	National Aquatic Resource Surveys
NAWQA	National Water-Quality Assessment
NBEP	Narragansett Bay Estuary Program
NEP	National Estuary Program
NFMR	North Fork Maquoketa River
NJ DEP	New Jersey Department of Environmental Protection
NOAA	National Oceanic and Atmospheric Administration
NPDES	National Pollutant Discharge Elimination System
NPS	nonpoint source
NTU	nephelometric turbidity unit
NWQC	numeric water quality criteria
O/E	observed over expected
ODEQ	Oregon Department of Environmental Quality
ODFW	Oregon Department of Fish and Wildlife
ONRW	Outstanding National Resource Water
ORD	Office of Research and Development (U.S. Environmental Protection Agency)
PA DEP	Pennsylvania Department of Environmental Protection
PREDATOR	PREDictive Assessment Tool for Oregon
QHEI	qualitative habitat evaluation index
RIVPACS	River Invertebrate Prediction and Classification System
SI	stressor identification
SSH	seasonal salmonid habitat
TBEP	Tampa Bay Estuary Program
TIE	toxicity identification evaluation
TMDL	Total Maximum Daily Load
TRE	toxicity reduction evaluation
TS	temperature stress
UAA	use attainability analysis
UCONN	University of Connecticut
USGS	U.S. Geological Survey
UT DEQ	Utah Department of Environmental Quality

VA DCR	Virginia Department of Conservation and Recreation
VSA	Virtual Stream Assessment
VT DEC	Vermont Department of Environmental Conservation
WET	whole effluent toxicity
WQL	water quality limited
WQS	water quality standards
WWH	warmwater habitat
WWTF	wastewater treatment facility
WWTP	wastewater treatment plant

APPENDIX A. ADDITIONAL RESOURCES

Biological Assessment and Biological Criteria: Technical Guidance

Biological assessment and biological criteria	**Description/summary**
Biological Criteria: National Program for Surface Waters (EPA 440-5-90-004) Source: U.S. Environmental Protection Agency Date of Publication: 1990	This document provides EPA regions, states and others with the conceptual framework and assistance necessary to develop and implement narrative and numeric biological criteria and to promote national consistency in application.
http://www.epa.gov/bioindicators/	
Policy on the Use of Bioassessments and Criteria in the Water Quality Program Source: U.S. Environmental Protection Agency Date of Publication: 1991	This document provides policy guidance on integration of biological surveys, assessments, and criteria with chemical-specific analysis and whole effluent and ambient toxicity testing methods in the water quality program.
http://www.epa.gov/bioiweb1/pdf/PolicyonBiologicalAssessmentsandCriteria.pdf	
Coral reefs	**Description/summary**
Stony Coral Rapid Bioassessment Protocol (EPA 600-R-06-167) Source: U.S. Environmental Protection Agency Date of Publication: 2007	The principal purpose of the *Stony Coral Rapid Bioassessment Protocol* is to introduce a simple and rapid coral survey method that provides multiple biological indicators to characterize coral condition.

Biological assessment and biological criteria	**Description/summary**
	The document offers insight on indicator relevance to ecosystem services (societal values), reef condition, and sustainability. It provides information regarding regulatory programs, and it presents a few examples describing how biological assessment indicators can be incorporated into a regulatory biological criteria program to conserve coral resources.
http://www.epa.gov/bioindicators/	
Coral Reef Biological Criteria: Using the Clean Water Act to Protect a National Treasure (EPA-600-R-10-054) Source: U.S. Environmental Protection Agency Date of Publication: 2010	Coral reef resource managers can use this document as a guide for developing and implementing biological criteria as part of water quality standards. Biological criteria are complementary to chemical and physical criteria and, once established, carry the same regulatory authority. The document introduces the role of biological criteria under the Clean Water Act and describes the process for identifying metrics, establishing reference values, designing a long-term monitoring program, and integrating biological criteria with existing management programs. It includes sections that link biological criteria to high-visibility issues such as ecosystem services, climate change, and ocean acidification.
http://cfpub.epa.gov/si/si public record report.cfm?dirEntryId=223392	
Estuaries and coastal waters	**Description/summary**
Estuarine and Coastal Marine Waters: Bioassessment and Biocriteria Technical Guidance (EPA 822-B-00-024) Source: U.S. Environmental Protection Agency Date of Publication: 2000	This technical guidance provides an extensive collection of methods and protocols for conducting biological assessments in estuarine and coastal marine waters and the procedures for deriving biological criteria from the results. See also *National Coastal Condition Reports* (2001, 2004 and 2008) under *National Aquatic Resource Surveys* listed below.

(Continued)

http://www.epa.gov/waterscience/biocriteria/States/estuaries/estuaries.pdf	
Lakes and reservoirs	**Description/summary**
Lakes and Reservoir Bioassessment and Biocriteria Technical Guidance Document (EPA 841-B-98-007) Source: U.S. Environmental Protection Agency Date of Publication: 1998	This guidance is intended to provide managers and field biologists with functional methods and approaches that will facilitate the implementation of viable lake biological assessment and biological criteria programs that meet their needs and resources. Procedures for program design, reference condition determination, field biological surveys, biological criteria development, and data analysis are detailed. In addition, the document provides information on the application and effectiveness of lake biological assessment to existing EPA and state/tribal programs such as the Clean Lakes Program, 305(b) assessments, NPDES permitting, risk assessment, and watershed management. See also *National Lakes Assessment Report* (2010) under *National Aquatic Resource Surveys* listed below.
http://www.epa.gov/owow/monitoring/tech/lakes.	
Non-wadeable streams and rivers	**Description/summary**
Concepts and Approaches for the Bioassessment of Non- wadeable Streams and Rivers (EPA 600-R-06-127) Source: U.S. Environmental Protection Agency Date of Publication: 2006	This document provides a framework for the development of biological assessment programs and biological criteria for large rivers.
Non-wadeable streams and rivers	**Description/summary**
	It helps states establish or refine their large river protocols for field sampling, laboratory sample processing, data management and analysis, and assessment and reporting.
http://www.epa.gov/eerd/rivers/non-wadeable full doc.pdf	

Streams and wadeable rivers	Description/summary
Biological Criteria: Technical Guidance for Streams and Small Rivers (EPA 822-B-96-001) Source: U.S. Environmental Protection Agency Date of Publication: 2001	The goal of this document is to help states develop and use biological criteria for streams and small rivers. It includes a general strategy for biological criteria development, identifies steps in the process, and provides technical guidance on how to complete each step, using the experience and knowledge of existing state, regional, and national surface water programs. See also *Wadeable Streams Assessment Report* (2006) under *National Aquatic Resource Surveys* listed below.
http://www.epa.gov/bioindicators/	
revisededition1996.pdf	
Rapid Bioassessment Protocols for Use in Streams and Wadeable Rivers: Periphyton, Benthic Macroinvertebrates and Fish, 2nd ed. (EPA 841-B-99-002) Source: U.S. Environmental Protection Agency Date of Publication: 1999	This document is a practical technical reference for conducting cost-effective biological assessments of lotic systems. The Rapid Bioassessment Protocols (RBPs) are a blend of existing methods used by various states to sample biological assemblages and assess physical habitat
http://www.epa.gov/owow/monitoring/rbp/download.html	

Other Relevant Water Program Guidance

Listing and TMDLs	Description/summary
Memorandum: Clarification of the Use of Biological Data and Information in the 2002 Integrated Water Quality Monitoring and Assessment Report Guidance	This memorandum modified the 2002 *Integrated Water Quality Monitoring and Assessment Report Guidance* to provide clarity and promote consistency in the
Listing and TMDLs	**Description/summary**
Source: U.S. Environmental Protection Agency Date of Publication: 2002	manner in which states use biological data and information in developing their submissions.
http://water.	

(Continued)

Listing and TMDLs	**Description/summary**
Guidance for 1994 Section 303(d) Lists Source: U.S. Environmental Protection Agency Date of Publication: 1994	This memorandum clarified how biological data can be used to support listing of a waterbody on the section 303(d) list.
http://water.	
Recovery Potential Screening Source: U.S. Environmental Protection Agency Date of Publication: 2012	The Recovery Potential Screening website is a user-driven, flexible approach for comparing relative differences in restorability among impaired waters. The screening process uses ecological, stressor, and social indicators to evaluate and compare waters and reveal factors that may explain the relative restorability of waters. This technical method and website are intended to assist in complex planning and prioritizing decisions, provide a systematic and transparent comparison approach, reveal underlying environmental and social factors that affect restorability, and better inform restoration strategies to help achieve results. The website provides step-by-step directions in the screening process, downloadable tools for calculating indices and displaying results, summaries of indicators and their measurement from common data sources, a recovery literature database, and several case studies and related links.
http://www.epa.gov/recoverypotential/	
Monitoring and assessment	Description/summary
Guidance for 2006 Assessment, Listing and Reporting Requirements Pursuant to Sections 303(d), 305(b) and 314 of the Clean Water Act Source: U.S. Environmental Protection Agency Date of Publication: 2005	This guidance is for states, territories, authorized tribes, and interstate commissions that help prepare and submit section 305(b) reports (referred to as *jurisdictions*). It outlines the development of biennial Integrated Reports, which that would support EPA's strategy for achieving a broad-scale, national inventory of water quality conditions.

Monitoring and assessment	Description/summary
	The objective of this guidance is to provide jurisdictions (1) a recommended reporting format and (2) suggested content to be used in developing a single document that integrates the reporting requirements of CWA sections 303(d), 305(b), and 314. (Pursuant to the CWA, jurisdictions report to EPA biannually on the condition of waters within their boundaries.)
http://www.epa.gov/owow/tmdl/2006IRG/report/2006irg-report.pdf	
Elements of a State Water Monitoring and Assessment Program (EPA 841-B-03-003) Source: U.S. Environmental Protection Agency Date of Publication: 2003	This document recommends 10 basic elements of a state water monitoring program and serves as a tool to help EPA and states determine whether a monitoring program meets the prerequisites of CWA section 106(e)(1).
http://www.epa.gov/owow/monitoring/elements/	
Consolidated Assessment and Listing Methodology (CALM): Toward a Compendium of Best Practices Source: U.S. Environmental Protection Agency Date of Publication: 2002	CALM provides a framework for states and other jurisdictions to document how they collect and use water quality data and information for environmental decision making. The primary purposes of the data analyses are to determine the extent to which all waters are attaining water quality standards, to identify waters that are impaired and need to be added to the 303(d) list, and to identify waters that can be removed from the list because they are attaining standards.
http://www.epa.gov/owow/monitoring/calm.html	
Biological Criteria: Technical Guidance for Survey Design and Statistical Evaluation of Biosurvey Data (EPA 822-B97-002) Source: U.S. Environmental Protection Agency Date of Publication: 1997	The emphasis of this guidance is on the practical application of basic statistical concepts to the development of biological criteria for surface water resource protection, restoration, and management.
http://www.epa.gov/bioindicators/	
TechnicalGuidanceforSurveyDesignandStatisticalEvaluationofBiosurveyData.pdf	
Generic Quality Assurance Project Plan Guidance for Programs Using Community Level Biological Assessment in Wadeable Streams and Rivers	This document represents generic guidance for development of QAPPs for specific biological assessment projects or programs.

(Continued)

Monitoring and assessment	**Description/summary**
(EPA 841-B-95-004) Source: U.S. Environmental Protection Agency Date of Publication: 1995	It has been specifically designed for use by states using biological assessment protocols that focus on community-level responses as indicated by a multimetric approach and taxonomy to the genus/species level.
http://www.epa.gov/bioindicators/	
National Aquatic Resource Surveys: *National Coastal Condition Report.* (2001) EPA-620/R-01/005 *National Coastal Condition Report II.* (2004) EPA-620/R- 03/002 *Wadeable Streams Assessment.* (2006) EPA-841-B-06-002 *National Coastal Condition Report III.* (2008) EPA/842-R-08- 002 *National Lakes Assessment.* (2010) EPA-841-R-09-001 Source: U.S. Environmental Protection Agency Dates of Publication: see above	The surveys are conducted using a statistical survey design to yield unbiased, statistically representative estimates of the biological condition of the whole water resource (e.g., wadeable streams, lakes, rivers). Data are collected, processed, and analyzed through EPA-state collaboration to assess and report on the condition of the nation's waters with documented confidence. Surveys collect a suite of indicators relating to the biological/physical habitat and water quality of the resource to assess the resource condition and determine the percentage meeting the goals of the CWA. Surveys collect information on biological and abiotic factors at 30–50 sites on an ecoregion level II scale for each resource.
http://www.epa.gov/owow/monitoring/nationalsurveys.html	
http://www.epa.gov/owow/oceans/	
http://www.epa.gov/owow/streamsurvey/	
http://www.epa.gov/owow/lakes/	
Predictive Tools	Description/summary
Landscape and Predictive Tools: A Guide to Spatial Analysis for Environmental Assessment (draft) (EPA-100-R-11-002) Source: U.S. Environmental Protection Agency Date of Publication: In process of finalization. Release expected 2012.	This methods manual describes the purpose, rationale, and basic steps for using landscape and predictive tools for Clean Water Act monitoring, assessment, and management purposes such as filling monitoring gaps and prioritizing protection and rehabilitation actions.

Predictive Tools	Description/summary
	This guidance stresses simultaneous use of matched (or paired) landscape and in situ data for empirical modeling to enhance predictive capabilities and encourage science-based targeting and priority setting. Example and potential applications include criteria and standards development, problem identification and prevention, prioritization and targeting of rehabilitation, and advancing science, education, and society's ability to effectively manage aquatic and terrestrial resources. This methods guidance is organized into four sections: (I) Introduction to Landscape and Predictive Tools; (II) Geographic Frameworks, Spatial Data, and Analysis Tools; (III) Examples and Case Studies; and (IV) Gaps and Needs for Research and Applications; plus an extensive Toolbox providing links to and short descriptions of a wide range of easily accessed data sets and analytical tools. Wider application of these tools and approaches should yield better protection for high-quality waters and quicker, more cost-effective restoration of impaired waters.
http://www.epa.gov/raf/pubecological.htm	
Stressor Response	
Causal Analysis/Diagnosis Decision Information System (CADDIS) Source: U.S. Environmental Protection Agency Date: Last updated September 23, 2010	The Causal Analysis/Diagnosis Decision Information System, or CADDIS, is a website developed to help scientists and engineers in the Regions, States, and Tribes conduct causal assessments in aquatic systems. It is organized into five volumes: Volume 1: Stressor Identification Volume 2: Sources, Stressors & Responses Volume 3: Examples & Applications Volume 4: Data Analysis Volume 5: Causal Databases
http://www.epa.gov/caddis	

(Continued)

Stressor Response	
Using Stressor-response Relationships to Derive Numeric Nutrient Criteria (EPA-820-2-10-001) Source: U.S. Environmental Protection Agency Date of Publication: 2010	This document provides guidance on statistical methods for estimating stressor-response relationships between changes in nutrient concentrations and changes in biological response variables. The document also provides guidance on methods for interpreting these relationships to derive numeric nutrient criteria. Other specific topics discussed include selecting appropriate covariates to improve the accuracy of estimated relationships, and methods for accounting for uncertainty in estimated relationships when deriving criteria.
http://water.	
Water quality-based toxics control	Description/summary
Technical Support Document for Water Quality-based Toxics Control (EPA-5052-90-001) Source: U.S. Environmental Protection Agency Date of Publication: 1991	This document provides technical guidance for assessing and regulating discharge of toxic substances to waters of the United States. It was issued in support of EPA regulations and policy initiatives involving the application of biological assessment and chemical techniques to control toxic pollution to surface waters.
http://www.epa.gov/npdes/pubs/owm0264.pdf	
Watershed Protection	Description/summary
Identifying and Protecting Healthy Watersheds: A Technical Guide (draft) Source: U.S. Environmental Protection Agency Date of Publication: In process of finalization. Release expected 2012.	This draft technical document provides an overview of the key concepts behind an approach to identify and protect healthy watersheds, examples of assessments of healthy watershed components, an integrated assessment framework for identifying healthy watersheds, examples of management approaches, sources of national data, and key assessment tools. It contains numerous examples and case studies from across the country. The intended audience for this document is aquatic resource scientists and managers at the state, tribal, regional, and local levels; non-governmental organizations; and federal agencies.

Watershed Protection	Description/summary
	It will also benefit local government land use managers and planners as they develop protection priorities.
http://water.	

End Notes

[1] http://water factsheet.pdf
[2] http://water
[3] http://water
[4] For an additional example, see http://water
[5] http://www.dnr.state.
[6] http://water
[7] http://www.epa.gov/roe.
[8] EPA Integrated Reporting Guidance for the 2006 Section 303(d) and 305(b) Reporting Cycle website: http://water index.cfm
[9] EPA Recovery Potential Screening website: http://www.epa.gov/recoverypotential.
[10] For more information on NDPES regulations, go to http://cfpub.epa.gov/npdes/regs.cfm?program id=45.
[11] As prescribed under NPDES regulatory requirements for permit reopeners/modifications (CFR 122.44). For more information on NDPES regulations, go to http://cfpub.epa.gov/npdes/regs.cfm?program id=45.
[12] Chesapeake Bay Program website: http://www.chesapeakebay.net.
[13] National Estuary Program website: http://water index.cfm.
[14] http://cfpub.epa.gov/caddis/index.cfm.
[15] More information about GIS is at http://www.gis.com/content/what-gis.
[16] Under the CWA, class B waters are designated for the protection of aquatic life uses. The WW-2 classification is for small streams.
[17] Anthropogenic stressors can be manifest changes in plant communities through displacement and competition from exotic invasive species. The FQI is the calculation of the plant communities' mean coefficient of conservatism multiplied by the square root of the number of species. The coefficient of conservatism is a measure of an individual species' fidelity to natural habitats and communities.
[18] For more information, see http://water

In: Use of Biological Assessments ... ISBN: 978-1-62948-481-5
Editor: Ian W. Sorenson

Chapter 2

CLEAN WATER ACT: A SUMMARY OF THE LAW[*]

Claudia Copeland

SUMMARY

The principal law governing pollution of the nation's surface waters is the Federal Water Pollution Control Act, or Clean Water Act. Originally enacted in 1948, it was totally revised by amendments in 1972 that gave the act its current dimensions. The 1972 legislation spelled out ambitious programs for water quality improvement that have since been expanded and are still being implemented by industries and municipalities.

This report presents a summary of the law, describing the statute without discussing its implementation.

The Clean Water Act consists of two major parts, one being the provisions which authorize federal financial assistance for municipal sewage treatment plant construction. The other is the regulatory requirements that apply to industrial and municipal dischargers. The act has been termed a technology-forcing statute because of the rigorous demands placed on those who are regulated by it to achieve higher and higher levels of pollution abatement under deadlines specified in the law. Early on, emphasis was on controlling discharges of conventional

[*] This is an edited, reformatted and augmented version of a Congressional Research Service publication, CRS Report for Congress RL30030, from www.crs.gov, prepared for Members and Committees of Congress, dated November 30, 2012.

pollutants (e.g., suspended solids or bacteria that are biodegradable and occur naturally in the aquatic environment), while control of toxic pollutant discharges has been a key focus of water quality programs more recently.

Prior to 1987, programs were primarily directed at point source pollution, that is, wastes discharged from discrete sources such as pipes and outfalls. Amendments to the law in that year authorized measures to address nonpoint source pollution (runoff from farm lands, forests, construction sites, and urban areas), which is estimated to represent more than 50% of the nation's remaining water pollution problems. The act also prohibits discharge of oil and hazardous substances into U.S. waters.

Under this act, federal jurisdiction is broad, particularly regarding establishment of national standards or effluent limitations. Certain responsibilities are delegated to the states, and the act embodies a philosophy of federal-state partnership in which the federal government sets the agenda and standards for pollution abatement, while states carry out day-to-day activities of implementation and enforcement.

To achieve its objectives, the act is based on the concept that all discharges into the nation's waters are unlawful, unless specifically authorized by a permit, which is the act's principal enforcement tool. The law has civil, criminal, and administrative enforcement provisions and also permits citizen suit enforcement.

Financial assistance for constructing municipal sewage treatment plants and certain other types of water quality improvements projects is authorized under title VI. It authorizes grants to capitalize State Water Pollution Control Revolving Funds, or loan programs. States contribute matching funds, and under the revolving loan fund concept, monies used for wastewater treatment construction are repaid to states, to be available for future construction in other communities.

INTRODUCTION

The principal law governing pollution of the nation's surface waters is the Federal Water Pollution Control Act, or Clean Water Act. Originally enacted in 1948, it was totally revised by amendments in 1972 that gave the act its current shape. The 1972 legislation spelled out ambitious programs for water quality improvement that have since been expanded and are still being implemented by industries, municipalities and others. Congress made fine-tuning amendments in 1977, revised portions of the law in 1981, and enacted further amendments in 1987.

This report presents a summary of the law, describing the statute. Many details and secondary provisions are omitted here, and even some major components are only briefly mentioned. Further, this report describes the statute, while other CRS products discuss implementation issues.[1] *Table 1* shows the original enactment and subsequent major amendments. *Table 2*, at the end of this report, cites the major U.S. Code sections of the codified statute.

Table 1. Clean Water Act and Major Amendments (Codified generally as 33 U.S.C. §§1251-1387)

Year	Act	Public Law
1948	Federal Water Pollution Control Act	P.L. 80-845
		(Act of June 30, 1948)
1956	Water Pollution Control Act of 1956	P.L. 84-660 (Act of July 9, 1956)
1961	Federal Water Pollution Control Act Amendments	P.L. 87-88
1965	Water Quality Act of 1965	P.L. 89-234
1966	Clean Water Restoration Act	P.L. 89-753
1970	Water Quality Improvement Act of 1970	P.L. 91-224, Part I
1972	Federal Water Pollution Control Act Amendments	P.L. 92-500
1977	Clean Water Act of 1977	P.L. 95-217
1981	Municipal Wastewater Treatment Construction Grants Amendments	P.L. 97-117
1987	Water Quality Act of 1987	P.L. 100-4

Authorizations for appropriations to support the law generally expired at the end of FY1990 (September 30, 1990). Programs did not lapse, however, and Congress has continued to appropriate funds to carry out the act. Since the 1987 amendments, although Congress has enacted several bills that reauthorize and modify a number of individual provisions in the law, none comprehensively addressed major programs or requirements.

BACKGROUND

The Federal Water Pollution Control Act of 1948 was the first comprehensive statement of federal interest in clean water programs, and it specifically provided state and local governments with technical assistance funds to address water pollution problems, including research. Water pollution was viewed as primarily a state and local problem, hence, there were no federally required goals, objectives, limits, or even guidelines. When it came to enforcement, federal involvement was strictly limited to matters involving interstate waters and only with the consent of the state in which the pollution originated.

During the latter half of the 1950s and well into the 1960s, water pollution control programs were shaped by four laws that amended the 1948 statute. They dealt largely with federal assistance to municipal dischargers and with federal enforcement programs for all dischargers. During this period, the federal role and federal jurisdiction were gradually extended to include navigable intrastate, as well as interstate, waters. Water quality standards became a feature of the law in 1965, requiring states to set standards for interstate waters that would be used to determine actual pollution levels and control requirements. By the late 1960s, there was a widespread perception that existing enforcement procedures were too time-consuming and that the water quality standards approach was flawed because of difficulties in linking a particular discharger to violations of stream quality standards. Additionally, there was mounting frustration over the slow pace of pollution cleanup efforts and a suspicion that control technologies were being developed but not applied to the problems. These perceptions and frustrations, along with increased public interest in environmental protection, set the stage for the 1972 amendments.

The 1972 statute did not continue the basic components of previous laws as much as it set up new ones. It set optimistic and ambitious goals, required all municipal and industrial wastewater to be treated before being discharged into waterways, increased federal assistance for municipal treatment plant construction, strengthened and streamlined enforcement, and expanded the federal role while retaining the responsibility of states for day-to-day implementation of the law.

The 1972 legislation declared as its objective the restoration and maintenance of the chemical, physical, and biological integrity of the nation's waters. Two goals also were established: zero discharge of pollutants by 1985 and, as an interim goal and where possible, water quality that is both "fishable" and "swimmable" by mid-1983. While those dates have passed, the goals remain, and efforts to attain them continue.

Overview

The Clean Water Act (CWA) today consists of two parts, broadly speaking, one being the Title II and Title VI provisions, which authorize federal financial assistance for municipal sewage treatment plant construction. The other is regulatory requirements, found throughout the act, that apply to industrial and municipal dischargers.

The act has been termed a technology-forcing statute because of the rigorous demands placed on those who are regulated by it to achieve higher and higher levels of pollution abatement. Industries were given until July 1, 1977, to install "best practicable control technology" (BPT) to clean up waste discharges. Municipal wastewater treatment plants were required to meet an equivalent goal, termed "secondary treatment," by that date. (Municipalities unable to achieve secondary treatment by that date were allowed to apply for case-by-case extensions up to July 1, 1988. According to EPA, 86% of all cities met the 1988 deadline; the remainder were put under administrative or court-ordered schedules requiring compliance as soon as possible. However, many cities continue to make investments in building or upgrading facilities needed to achieve secondary treatment, and funding needs remain high; see discussion below.) Cities that discharge wastes into marine waters were eligible for case-by-case waivers of the secondary treatment requirement, where sufficient showing could be made that natural factors provide significant elimination of traditional forms of pollution and that both balanced populations of fish, shellfish, and wildlife and water quality standards would be protected.

The primary focus of BPT was on controlling discharges of conventional pollutants, such as suspended solids, biochemical oxygen demanding materials, fecal coliform and bacteria, and pH. These pollutants are substances that are biodegradable (i.e., bacteria can break them down), occur naturally in the aquatic environment, and deplete the dissolved oxygen concentration in water, which is necessary for fish and other aquatic life.

The act also mandated greater pollutant cleanup than BPT by no later than March 31, 1989, generally requiring that industry use the "best available technology" (BAT) that is economically achievable. BAT level controls generally focus on toxic substances. Compliance extensions of as long as two years are available for industrial sources utilizing innovative or alternative technology. Failure to meet statutory deadlines could lead to enforcement action.

The CWA utilizes both water quality standards and technology-based effluent limitations to protect water quality. Technology-based effluent limitations are specific numerical limitations established by EPA and placed on certain pollutants from certain sources. They are applied to industrial and municipal sources through numerical effluent limitations in discharge permits issued by states or EPA (see discussion of "Permits, Regulations, and Enforcement," below). Water quality standards are standards for the overall quality of water. They consist of the designated beneficial use or uses of a waterbody (recreation, water supply, industrial, or other), plus a numerical or narrative statement identifying maximum concentrations of various pollutants that would not interfere with the designated use. The act requires each state to establish water quality standards for all bodies of water in the state. These standards serve as the backup to federally set technology-based requirements by indicating where additional pollutant controls are needed to achieve the overall goals of the act. In waters where industrial and municipal sources have achieved technology-based effluent limitations, yet water quality standards have not been met, dischargers may be required to meet additional pollution control requirements. For each of these waters, the act requires states to set a total maximum daily load (TMDL) of pollutants at a level that ensures that applicable water quality standards can be attained and maintained. A TMDL is both a planning process for attaining water quality standards and a quantitative assessment of pollution problems, sources, and pollutant reductions needed to restore and protect a river, stream, or lake. Based on state reports, EPA estimates that more than 41,000 U.S. waters are impaired and require preparation of TMDLs.

Control of toxic pollutant discharges has been a key focus of water quality programs. In addition to the BPT and BAT national standards, states are required to implement control strategies for waters expected to remain polluted by toxic chemicals even after industrial dischargers have installed the best available cleanup technologies required under the law. Development of management programs for these post-BAT pollutant problems was a prominent element in the 1987 amendments and is a key continuing aspect of CWA implementation.

Prior to the 1987 amendments, programs in the Clean Water Act were primarily directed at point source pollution, wastes discharged from discrete and identifiable sources, such as pipes and other outfalls. In contrast, except for general planning activities, little attention had been given to nonpoint source pollution (runoff of stormwater or snowmelt from agricultural lands, forests, construction sites, and urban areas), despite estimates that it represents more than 50% of the nation's remaining water pollution problems. As it travels across land surface towards rivers and streams, rainfall and snowmelt runoff picks up pollutants, including sediments, toxic materials, and conventional wastes (e.g., nutrients) that can degrade water quality.

The 1987 amendments authorized measures to address such pollution by directing states to develop and implement nonpoint pollution management programs (Section 319 of the act). States were encouraged to pursue groundwater protection activities as part of their overall nonpoint pollution control efforts. Federal financial assistance was authorized to support demonstration projects and actual control activities. These grants may cover up to 60% of program implementation costs.

The CWA provides for special regulation of the discharge of oil or hazardous substances, because of the potentially catastrophic effects of such events on public health and welfare. Section 311 prohibits the discharge of oil or hazardous substances into U.S. waters. It also requires higher standards of care in the management and movement of oil, including a requirement for spill prevention plans; it enables the government to recover the costs of cleaning up oil and hazardous substance discharges; and it provides for penalties for such discharges. In 1990, Congress enacted the Oil Pollution Act, which partially amended Section 311 and established a comprehensive system for the cleanup of oil spills, adding a mechanism to impose liability for such spills.[2]

While the act imposes great technological demands, it also recognizes the need for comprehensive research on water quality problems. This is provided throughout the statute, on topics including pollution in the Great Lakes and Chesapeake Bay, in-place toxic pollutants in harbors and navigable waterways, and water pollution resulting from mine drainage. The act also authorizes support to train personnel who operate and maintain wastewater treatment facilities.

Federal and State Responsibilities

Under this act, federal jurisdiction is broad, particularly regarding establishment of national standards or effluent limitations. The EPA issues regulations containing the BPT and BAT effluent standards applicable to categories of industrial sources (such as iron and steel manufacturing, organic chemical manufacturing, petroleum refining, and others). Certain responsibilities can be assumed by qualified states, in lieu of EPA, and this act, like other environmental laws, embodies a philosophy of federal-state partnership in which the federal government sets the agenda and standards for pollution abatement, while states carry out day-to-day activities of implementation and enforcement. Responsibilities under the act that may be carried out by qualified states include authority to issue discharge permits to industries and municipalities and to enforce permits; 46 states have been authorized to administer this permit program. EPA issues discharge permits in the remaining states—Idaho, Massachusetts, New Hampshire, New Mexico—the District of Columbia, and most U.S. territories. In addition, as noted above, all states are responsible for establishing water quality standards.

Titles II and VI—Municipal Wastewater Treatment Construction

Federal law has authorized grants for planning, design, and construction of municipal sewage treatment facilities since 1956 (Act of July 9, 1956, or P.L. 84-660). Congress greatly expanded this grant program in 1972 in order to assist cities in meeting the act's pollution control requirements. Since that time Congress has authorized $65 billion and appropriated more than $88 billion in CWA funds to aid wastewater treatment plant construction. Grants are allocated among the states according to a complex statutory formula that combines two factors: state population and an estimate of municipal sewage treatment funding needs derived from a survey conducted periodically by EPA and the states. The most recent estimate indicated that, as of January 2008, $298 billion more would be required to build and upgrade municipal wastewater treatment plants in the United States and for other types of water quality improvement projects that are eligible for funding under the act, a 17% increase over the previous estimate from four years earlier.

Under the Title II construction grants program established in 1972, federal grants were made for several types of projects based on a priority list established by the states. Grants were generally available for as much as 55% of total project costs. For projects using innovative or alternative technology (such as reuse or recycling of water), as much as 75% federal funding was allowed. Recipients were responsible for non-federal costs but were not required to repay federal grants.

Policymakers have debated the balance between assisting municipal funding needs, which remain large, and the impact of aid programs such as the Clean Water Act's on federal spending and budget deficits. In the 1987 amendments, Congress balanced these competing priorities by extending federal aid for wastewater treatment construction through FY1994, yet providing a transition towards full state and local government responsibility for financing after that date. Grants under the previous Title II program were authorized through FY1990. Under Title VI of the act, grants to capitalize State Water Pollution Control Revolving Funds, or loan programs, were authorized beginning in FY1989 to replace the Title II grants. States contribute matching funds, and under the revolving loan fund concept, monies used for wastewater treatment construction will be repaid to a state fund, to be available for future construction in other communities.

All states now have functioning loan programs, but the shift from federal grants to loans has been easier for some than others. The new financing requirements have been a problem for cities (especially small towns) that have difficulty repaying project loans. Statutory authorization for grants to capitalize state loan programs expired in 1994; however, Congress has continued to provide annual appropriations. An issue affecting some cities is overflow discharges of inadequately treated wastes from municipal sewers and how cities will pay for costly remediation projects. In 2000, Congress amended the act to authorize a two-year $1.5 billion grant program to help cities reduce these wet weather flows. Authorization for that wet weather grant program expired at the end of FY2003 and has not been renewed.

PERMITS, REGULATIONS, AND ENFORCEMENT

To achieve its objectives, the CWA embodies the concept that all discharges into the nation's waters are unlawful, unless specifically authorized by a permit. Thus, more than 65,000 conventional industrial and municipal dischargers must obtain permits from EPA (or qualified states) under the act's National Pollutant Discharge Elimination System (NPDES) program (authorized in Section 402 of the act). NPDES permits also are required for more than 150,000 industrial and municipal sources of stormwater discharges. An NPDES permit requires the discharger (source) to attain technology-based effluent limits (BPT or BAT for industry, secondary treatment for municipalities, or more stringent where needed for water quality protection). Permits specify the effluent limitations a discharger must meet, and the deadline for compliance. Sources also are required to maintain records and to carry out effluent monitoring activities. Permits are issued for no more than five years and must be renewed thereafter to allow continued discharge.

The NPDES permit incorporates numerical effluent limitations issued by EPA. The initial BPT limitations focused on regulating discharges of conventional pollutants, such as bacteria and oxygen-consuming materials. The more stringent BAT limitations emphasize controlling toxic pollutants—

heavy metals, pesticides, and other organic chemicals. In addition to these limitations applicable to categories of industry, EPA has issued water quality criteria for more than 115 pollutants, including 65 named classes or categories of toxic chemicals, or "priority pollutants." These criteria recommend ambient, or overall, concentration levels for the pollutants and provide guidance to states for establishing water quality standards that will achieve the goals of the act.

A separate type of permit is required to dispose of dredged or fill material in the nation's waters, including wetlands. Authorized by Section 404 of the act, this permit program is administered by the U.S. Army Corps of Engineers, subject to and using EPA's environmental guidance. Some types of activities are exempt from permit requirements, including certain farming, ranching, and forestry practices which do not alter the use or character of the land; some construction and maintenance; and activities already regulated by states under other provisions of the act. EPA may delegate certain Section 404 permitting responsibility to qualified states and has done so twice (Michigan and New Jersey). For some time, the act's wetlands permit program has been one of the most controversial parts of the law. Some who wish to develop wetlands maintain that federal regulation intrudes on and impedes private land-use decisions, while environmentalists seek more protection for remaining wetlands and limits on activities that are authorized to take place in wetlands.

Nonpoint sources of pollution, which EPA and states believe are responsible for the majority of water quality impairments in the nation, are not subject to CWA permits or other regulatory requirements under federal law. They are covered by state programs for the management of runoff, under Section 319 of the act.

Other EPA regulations under the CWA include guidelines on using and disposing of sewage sludge and guidelines for discharging pollutants from land-based sources into the ocean. (A related law, the Ocean Dumping Act, 33 U.S.C. §§1401-45, regulates the intentional disposal of wastes into ocean waters.[3]) EPA also provides guidance on technologies that will achieve BPT, BAT, and other effluent limitations.

The NPDES permit, containing effluent limitations on what may be discharged by a source, is the act's principal enforcement tool. EPA may issue a compliance order or bring a civil suit in U.S. district court against persons who violate the terms of a permit. The penalty for such a violation can be as much as $25,000 per day. Stiffer penalties are authorized for criminal violations of the act—for negligent or knowing violations—of as much as $50,000 per day, three years' imprisonment, or both. A fine of as much as

$250,000, 15 years in prison, or both, is authorized for "knowing endangerment"—violations that knowingly place another person in imminent danger of death or serious bodily injury. Finally, EPA is authorized to assess civil penalties administratively for certain well-documented violations of the law. These civil and criminal enforcement provisions are contained in Section 309 of the act. EPA, working with the Army Corps of Engineers, also has responsibility for enforcing against entities who fail to obtain or comply with a Section 404 permit.

While the CWA addresses federal enforcement, the majority of actions taken to enforce the law are undertaken by states, both because states issue the majority of permits to dischargers and because the federal government lacks the resources for day-to-day monitoring and enforcement. Like most other federal environmental laws, CWA enforcement is shared by EPA and states, with states having primary responsibility. However, EPA has oversight of state enforcement and retains the right to bring a direct action where it believes that a state has failed to take timely and appropriate action or where a state or local agency requests EPA involvement. Finally, the federal government acts to enforce against criminal violations of the federal law.

In addition, individuals may bring a citizen suit in U.S. district court against persons who violate a prescribed effluent standard or limitation or permit requirement. Citizens also may bring suit against the Administrator of EPA for failure to carry out a nondiscretionary duty under the act.

SELECTED REFERENCES

Hamilton, Pixie, Timothy Miller, and Donna Myers. "Water Quality in the Nation's Streams and Aquifers—Overview of Selected Findings, 1991-2001." U.S. Geological Survey Circular 1265. May 2004.

Lavelle, Marianne. "Water Woes." U.S. News & World Report. June 4, 2007. pp. 37-53.

U.S. Congressional Budget Office. Public Spending on Transportation and Water Infrastructure. November 2010. Pub. No. 4088. 49 p.

U.S. Environmental Protection Agency. EPA's Draft Report on the Environment, Technical Document. Chapter 2, Purer Water. June 2003. EPA 600-R-03-050. pp. 2-1 - 2-64.

——. Clean Watersheds Needs Survey 2008, Report to Congress. May 2010. EPA-832-R-10-002.

—— Clean Water State Revolving Fund Programs 2009 Annual Report. June 2010. EPA-832-R- 10-001. 36 p.
—— National Water Quality Inventory: Report to Congress for the 2004 Reporting Cycle. January 2009. EPA-841-R-08-001. 37 p.
U.S. Government Accountability Office. Water Quality, A Catalog of Related Federal Programs. GAO/RCED-96-173. June 1996. 64 p.
—— Water Infrastructure: Information on Financing, Capital Planning, and Privatization. GAO-02-764. August 2002. 79 p.

Table 2. Major U.S. Code Sections of the Clean Water Act (codified generally as 33 U.S.C., Chapter 26, Sections 1251-1387)

33 U.S.C.	Section Title	Clean Water Act (as amended)
Subchapter I -	Research and Related Programs	
1251	Declaration of goals and policy	Sec. 101
1252	Comprehensive programs for water pollution control	Sec. 102
1253	Interstate cooperation and uniform laws	Sec. 103
1254	Research, investigations, training and information	Sec. 104
1255	Grants for research and development	Sec. 105
1256	Grants for pollution control programs	Sec. 106
1257	Mine water pollution control demonstrations	Sec. 107
1258	Pollution control in the Great Lakes	Sec. 108
1259	Training grants and contracts	Sec. 109
1260	Applications for training grants or contracts; allocations of grants or contracts	Sec. 110
1261	Award of scholarships	Sec. 111
1262	Definitions and authorizations	Sec. 112
1263	Alaska village demonstration project	Sec. 113
1264	Lake Tahoe study	Sec. 114
1265	In-place toxic pollutants	Sec. 115
1266	Hudson River PCB reclamation demonstration project	Sec. 116
1267	Chesapeake Bay	Sec. 117
1268	Great Lakes	Sec. 118
1269	Long Island Sound	Sec. 119
1270	Lake Champlain Basin program	Sec. 120
1273	Lake Pontchartrain Basin	Sec. 121

33 U.S.C.	Section Title	Clean Water Act (as amended)
1274	Wet weather watershed pilot projects	Sec. 122
Subchapter II -	Grants for Construction of Treatment Works	
1281	Purpose	Sec. 201
1282	Federal share	Sec. 202
1283	Plans, specifications, estimates, and payments	Sec. 203
1284	Limitations and conditions	Sec. 204
1285	Allotment	Sec. 205
1286	Reimbursement and advanced construction	Sec. 206
1287	Authorization	Sec. 207
1288	Areawide waste treatment management	Sec. 208
1289	Basin planning	Sec. 209
1290	Annual survey	Sec. 210
1291	Sewage collection systems	Sec. 211
1292	Definitions	Sec. 212
1293	Loan guarantees for construction of treatment works	Sec. 213
1294	Public information on water recycling, reuse	Sec. 214
1295	Requirements for American materials	Sec. 215
1296	Determination of priority	Sec. 216
1297	Cost-effectiveness guidelines	Sec. 217
1298	Cost effectiveness	Sec. 218
1299	State certification of projects	Sec. 219
1300	Pilot program for alternative water source projects	Sec. 220
1301	Sewer overflow control grants	Sec. 221
Subchapter III -	Standards and Enforcement	
1311	Effluent Limitations	Sec. 301
1312	Water quality related effluent limitations	Sec. 302
1313	Water quality standards and implementation plans	Sec. 303
1314	Information and guidelines	Sec. 304
1315	Water quality inventory	Sec. 305
1316	National standards of performance	Sec. 306
1317	Toxic and pretreatment effluent standards	Sec. 307
1318	Inspections, monitoring, and entry	Sec. 308
1319	Federal enforcement	Sec. 309

Table 2. (Continued)

33 U.S.C.	Section Title	Clean Water Act (as amended)
1320	International pollution abatement	Sec. 310
1321	Oil and hazardous substance liability	Sec. 311
1322	Marine sanitation devices	Sec. 312
1323	Federal facilities pollution control	Sec. 313
1324	Clean lakes	Sec. 314
1325	National study commission	Sec. 315
1326	Thermal discharges	Sec. 316
1327	Financing study	Sec. 317
1328	Aquaculture	Sec. 318
1329	Nonpoint source management programs	Sec. 319
1330	National estuary program	Sec. 320
Subchapter IV -	Permits and Licenses	
1341	Certification	Sec. 401
1342	National pollutant discharge elimination system	Sec. 402
1343	Ocean discharge criteria	Sec. 403
1344	Permits for dredged or fill material	Sec. 404
1345	Disposal of sewage sludge	Sec. 405
1346	Coastal recreation water quality monitoring and notification	Sec. 406
Subchapter V -	General Provisions	
1361	Administration	Sec. 501
1362	General definitions	Sec. 502
1363	Water pollution control advisory board	Sec. 503
1364	Emergency powers	Sec. 504
1365	Citizen suits	Sec. 505
1366	Appearance	Sec. 506
1367	Employee protection	Sec. 507
1368	Federal procurement	Sec. 508
1369	Administrative procedure and judicial review	Sec. 509
1370	State authority	Sec. 510
1371	Other affected authority	Sec. 511
1372	Separability	Sec. 512
1372	Labor standards	Sec. 513
1373	Public health agency coordination	Sec. 514
1374	Effluent standards and water quality information advisory committee	Sec. 515

33 U.S.C.	Section Title	Clean Water Act (as amended)
1375	Reports to Congress	Sec. 516
1376	General authorization	Sec. 517
1377	Indian tribes	Sec. 518
1251 note	Short title	Sec. 519
Subchapter VI -	State Water Pollution Control Revolving Funds	
1381	Grants to states for establishment of revolving funds	Sec. 601
1382	Capitalization grant agreements	Sec. 602
1383	Water pollution control revolving loan funds	Sec. 603
1384	Allotment of funds	Sec. 604
1385	Corrective action	Sec. 605
1386	Audits, reports, and fiscal controls; intended use plan	Sec. 606
1387	Authorization of appropriations	Sec. 607

Note: This table shows only the major code sections. For more detail and to determine when a section was added, the reader should consult the official printed version of the U.S. Code.

End Notes

[1] For example, see CRS Report R41594, Water Quality Issues in the 112th Congress: Oversight and Implementation, by Claudia Copeland.

[2] P.L. 101-380; 33 U.S.C. § 2701 et seq.

[3] CRS Report RS20028, Ocean Dumping Act: A Summary of the Law, by Claudia Copeland.

INDEX

#

20th century, 31

A

abatement, ix, x, 109, 123, 124, 127, 129, 136
access, 5, 35, 64, 81, 82
accounting, 120
advancements, 1, 2, 39
adverse effects, 28, 69
advocacy, 80
agencies, 10, 16, 39, 75, 78, 81, 93, 96, 120
agriculture, 11, 31
air temperature, 58
Alaska, 134
algae, 107
alternative water sources, 64
alters, 105
ammonia, 12, 36, 37, 60, 62, 63, 64
amphibians, viii, ix, 3, 4, 27
appropriations, 125, 131, 137
aquatic biota, vii, 2, 5, 6, 7, 8, 11, 15, 25, 28, 31, 38, 45, 51, 53, 74, 75, 80, 84
aquatic habitats, 27, 65, 78
aquatic systems, 35, 36, 119
ARS, 46
assessment tools, 16, 89, 91, 120
assets, 99
atmosphere, 85
authority(s), 15, 113, 129, 136
awareness, 81, 98

B

backwaters, 92
bacteria, ix, 41, 107, 124, 127, 131
banks, 73, 74
base, 14, 33
benchmarks, 16, 57, 59, 60, 95
benefits, 18, 38
benthic invertebrates, 47, 48
benthic macroinvertebrates, viii, 3, 22, 52, 58, 61, 72, 75, 77, 79, 104
BI, 70
bias, 84, 85
bioindicators, 101, 112, 113, 115, 117, 118
biological assessments, vii, viii, 1, 2, 3, 4, 5, 6, 7, 8, 9, 10, 11, 12, 13, 14, 15, 16, 18, 19, 36, 49, 60, 61, 63, 67, 68, 74, 76, 78, 81, 86, 87, 88, 89, 92, 113, 115
biological criteria, vii, 2, 7, 9, 10, 12, 14, 15, 16, 17, 19, 22, 36, 39, 40, 41, 43, 44, 45, 46, 47, 54, 56, 57, 61, 65, 78, 79, 80, 84, 87, 104, 107, 109, 112, 113, 114, 115, 117
biological processes, 89
biological responses, 91
biomass, 30, 100

biomonitoring, 104
biotic, viii, ix, 3, 4, 26, 29, 37, 47, 48, 50, 51, 77, 88, 104, 106, 110
blueprint, 22
budget deficit, 130
Bureau of Land Management, 97
businesses, 60, 64

C

calibration, 29, 30
candidates, 82
carbon, 85
Caribbean, 97, 100, 101
case study(s), 9, 10, 13, 15, 16, 21, 24, 25, 27, 34, 39, 41, 116, 120
cattle, 64
causation, 37
certification, 79, 135
CFR, 12, 14, 121
challenges, vii, viii, 3
chemical(s), vii, viii, ix, 2, 3, 4, 5, 8, 9, 10, 11, 12, 14, 15, 17, 20, 24, 37, 38, 41, 45, 47, 49, 55, 56, 62, 64, 68, 70, 71, 72, 73, 78, 83, 105, 106, 107, 108, 109, 112, 113, 120, 126, 128, 129, 132
chlorination, 69
chlorine, 69
city(s), 101, 127, 130, 131
clarity, 115
classes, 10, 20, 41, 58, 79, 80, 94, 95, 132
classification, 20, 22, 40, 43, 44, 54, 55, 79, 90, 121
Clean Water Act (CWA), v, vii, viii, ix, 1, 2, 4, 41, 52, 61, 78, 97, 105, 110, 113, 116, 118, 123, 124, 125, 126, 128, 130, 134, 136
cleaning, 129
cleanup, 15, 126, 127, 128, 129
climate, vii, viii, 3, 4, 8, 29, 84, 85, 86, 91, 105, 113
climate change, vii, viii, 3, 4, 8, 85, 91, 113
climatic shifts, 84
coastal ecosystems, 16
collaboration, 10, 11, 81, 118
commercial, 31
communication, 17
communities, x, 6, 11, 36, 44, 45, 50, 66, 71, 72, 74, 76, 77, 79, 81, 84, 85, 86, 87, 89, 91, 92, 100, 104, 107, 108, 109, 121, 124, 131
competition, 109, 121
complement, ix, 4, 8, 25
compliance, 14, 36, 68, 69, 127, 131, 132
composition, 3, 7, 10, 26, 28, 31, 45, 46, 51, 77, 81, 85, 86, 87, 104
conceptual model, 17, 97
conflict, 2
Congress, 105, 123, 124, 125, 129, 130, 131, 133, 134, 137
Congressional Budget Office, 133
connectivity, 26, 100
consensus, 31, 67
consent, 126
conservation, 81, 83
construction, ix, x, 123, 124, 126, 127, 128, 130, 132, 135
controversial, 132
cooperation, 134
coordination, 136
copper, 68, 69
coral reefs, 86, 87, 90
correlation, 52
cost, 53, 60, 66, 96, 115, 119
crop, 64
crop residue, 64
CT, 22, 66, 110
cultivation, 31
CWA, viii, 3, 9, 12, 13, 14, 16, 17, 19, 24, 55, 79, 110, 117, 118, 121, 127, 128, 129, 130, 131, 132, 133
cycling, 28

D

danger, 133
data analysis, 19, 30, 114
data collection, 20, 34
data set, 8, 30, 37, 44, 45, 119
database, 18, 20, 45, 46, 80, 81, 116

decision makers, 2
decomposition, 28
Deer, 60
degradation, 4, 6, 14, 31, 36, 77, 78
demonstrations, 134
Department of the Interior, 97
deposition, 37, 74, 75
detectable, 84
detection, 31, 77
detention, 75
direct action, 133
direct measure, viii, 3, 8, 57
discharges, ix, x, 9, 14, 42, 47, 48, 65, 85, 101, 123, 124, 127, 128, 129, 131, 136
discriminant analysis, 50, 53
disinfection, 69
displacement, 121
dissolved oxygen, 36, 37, 41, 56, 62, 72, 127
distribution, 6, 10, 27, 45, 83, 91, 94, 108
District of Columbia, 15, 33, 39, 76, 77, 130
diversity, 3, 61, 65, 87, 104
dominance, 28, 31, 51, 68
draft, 2, 16, 89, 94, 97, 103, 118, 120
drainage, 55, 65, 82, 83, 129
drought, 29, 108
dumping, 5, 15

E

ecological information, 40
ecological processes, 86
ecosystem, viii, 3, 5, 11, 16, 26, 27, 28, 31, 53, 88, 90, 91, 100, 104, 105, 107, 113
education, 119
effluent(s), ix, x, 4, 8, 14, 15, 36, 38, 42, 45, 67, 68, 69, 70, 109, 112, 124, 127, 129, 131, 132, 133, 135
endangered, 5, 43, 108
enforcement, x, 15, 34, 76, 78, 124, 126, 127, 129, 132, 133, 135
engineering, 66
environment(s), ix, 3, 19, 87, 88, 106, 108, 124, 127
environmental characteristics, 60
environmental conditions, 26, 58, 104, 105, 109
environmental protection, 91, 126
Environmental Protection Agency, 1, 2, 6, 36, 51, 84, 86, 87, 89, 97, 98, 100, 101, 102, 103, 110, 111, 112, 113, 114, 115, 116, 117, 118, 119, 120, 133
environmental quality, 5
environmental variables, 95
erosion, 49, 74, 75, 76
evidence, 8, 12, 34, 37, 78, 93, 96
exclusion, 80
expertise, 28
exposure, ix, 4, 5, 37, 93, 109
extraction, 71

F

farm land, ix, 124
fauna, 31
federal aid, 130
federal assistance, 126
Federal Geographic Data Committee, 90
federal government, x, 124, 129, 133
federal law, 132, 133
Federal Register, 101
fidelity, 121
financial, ix, 123, 127, 129
fish, viii, ix, 3, 4, 5, 8, 9, 10, 11, 15, 22, 26, 28, 34, 36, 37, 41, 43, 47, 48, 51, 52, 54, 55, 60, 61, 65, 66, 71, 74, 75, 76, 77, 79, 81, 82, 88, 97, 100, 104, 110, 127
fisheries, 55, 65, 100
flooding, 45, 75
fluctuations, 62
food, 27, 37, 50, 108, 109
food web, 50
formula, 130
freshwater, 5, 41, 92, 99
funding, 127, 130
funds, x, 66, 124, 125, 130, 131, 137

G

GAO, 134
genus, 108, 118
GIS, 20, 48, 92, 110, 121
global climate change, 8, 85
goal-setting, 26
GPS, 82, 110
grants, x, 82, 124, 129, 130, 131, 134, 135
grass, 88
groundwater, 28, 129
group membership, 86
grouping, 108
growth, 105, 106, 109
guidance, 2, 17, 36, 49, 67, 83, 112, 113, 114, 115, 116, 117, 119, 120, 132
guidelines, 61, 62, 125, 132, 135
Gulf of Mexico, 11

H

habitat quality, 56, 75, 79
habitats, 27, 51, 87, 88, 89, 92, 100, 108
harbors, 129
hazardous substances, x, 124, 129
health, 16, 28, 50, 57, 72, 82, 83, 91, 93, 104, 136
heavy metals, 36, 132
historical data, 86, 91, 106
history, 27, 41, 88, 108
host, 83
human, 11, 45, 84, 92, 106, 107
human activity, 107
human health, 106
Hurricane Katrina, 11
hydrologic regime, 42, 67
hypothesis, 36
hypoxia, 37, 91

I

ID, 34
ideal, 82, 107
identification, 8, 12, 15, 19, 33, 34, 35, 44, 57, 58, 60, 62, 64, 65, 68, 81, 109, 110, 111, 119
impact assessment, 10
impairments, 15, 19, 25, 36, 50, 54, 63, 64, 65, 67, 103, 107, 132
imprisonment, 132
improvements, x, 5, 6, 8, 10, 11, 15, 16, 22, 23, 33, 36, 38, 43, 69, 71, 73, 76, 124
individuals, 80, 81, 133
industry(s), ix, 43, 123, 124, 127, 130, 131, 132
infrastructure, 4
injury, 133
insects, 8, 73
integration, 19, 112
integrators, 86
integrity, vii, viii, 2, 3, 5, 7, 9, 11, 44, 45, 46, 47, 48, 50, 62, 71, 77, 80, 81, 82, 87, 88, 98, 104, 105, 106, 107, 110, 111, 126
invertebrates, 55, 77
investment(s), 4, 11, 127
Iowa, 12, 37, 39, 60, 61, 62, 63, 98, 99, 110
iron, 129
issues, 20, 54, 113, 125, 129

J

jurisdiction, x, 83, 124, 126, 129
justification, 49

K

kill, 15, 34, 61, 76, 77

L

Lake Pontchartrain, 134
lakes, 4, 24, 28, 35, 92, 114, 118, 136
landscape(s), 17, 20, 33, 38, 50, 51, 86, 87, 89, 90, 91, 92, 118, 119
laws, 126, 129, 133, 134
lead, 34, 82, 127
legislation, ix, 123, 124, 126

lesions, ix, 4, 28, 51
levees, 28
life cycle, 3
limestone, 50
loans, 131
local conditions, 28, 30, 83
local government, 121, 125, 130
Luxemburg, 61

M

macrobenthos, 104
majority, 46, 51, 59, 132, 133
mangrove forests, 88, 90
mangroves, 87
manufacturing, 129
manure, 60, 64
mapping, 80, 82
Maryland, 9, 10, 15, 33, 39, 47, 48, 49, 76, 77, 110
MAS, 100
materials, 28, 48, 127, 128, 131, 135
measurement(s), 59, 72, 107, 116
membership, 86
metals, 4, 15, 54, 68, 91, 132
meter, 81
methodology, 12, 19, 37
Mexico, 130
Miami, 36
micrograms, 68
microhabitats, 107
migration, 28, 95
Mississippi River, 11
models, viii, 3, 21, 57, 58, 59, 60, 86, 92, 107
modifications, 44, 55, 56, 79, 121
morphology, 31
mosaic, 90
Moses, 99

N

naming, 5
National Park Service, 77
native population, 58
native species, ix, 4, 31, 45, 52, 88, 106
natural habitats, 3, 87, 104, 121
natural resources, 82
nitrite, 62
nitrogen, vii, viii, 3, 4, 11, 34, 62, 68, 89, 90, 91
NOAA, 88, 92, 94, 96, 99, 111
North America, 5, 28, 97, 99
NPS, 5, 8, 14, 15, 39, 58, 59, 60, 66, 74, 95, 111
nutrient(s), vii, viii, 3, 5, 12, 28, 36, 54, 60, 63, 68, 82, 85, 91, 92, 120, 128
nutrient concentrations, 120

O

oceans, 118
OH, 100
oil, x, 11, 124, 129
oil spill, 11, 129
operations, 67, 69, 72
opportunities, 19, 67
organ, 87
organic chemicals, 132
organism, 26, 29, 87, 104, 107, 109
organize, 35
oversight, 79, 133
oxygen, 36, 42, 43, 56, 62, 64, 72, 73, 127, 131

P

Pacific, 94
pathogens, 4
pathways, 37
peer review, 22, 30
penalties, 129, 132
percentile, 45, 46, 47
periphyton, viii, 3, 107
permit, x, 1, 10, 14, 42, 48, 54, 68, 69, 78, 79, 121, 124, 130, 131, 132, 133
pesticide, 15, 34, 76, 77
petroleum, 129

pH, 49, 62, 69, 127
phosphorous, 89
phosphorus, vii, viii, 3, 4, 11, 35, 36, 60, 62, 63, 64, 68, 69, 70, 89, 90
physical structure, 90, 105
phytoplankton, 28
plants, x, 105, 124, 127, 130
policy, viii, 2, 4, 40, 41, 42, 43, 48, 109, 112, 120, 134
policy initiative, 120
pollutant source controls, vii, 2
pollutants, ix, 8, 13, 14, 42, 55, 60, 63, 64, 65, 68, 110, 124, 126, 127, 128, 129, 131, 132, 134
ponds, 73
pools, 56
population, 37, 61, 87, 93, 106, 108, 109, 130
population density, 108
precipitation, 29, 58
predictor variables, 85
preparation, 128
prevention, 119, 129
probability, 6
procurement, 136
project, 49, 66, 75, 82, 83, 91, 92, 96, 130, 131, 134
propagation, 9, 88
protection, viii, 3, 9, 11, 16, 38, 40, 41, 44, 47, 48, 50, 53, 55, 61, 81, 82, 90, 93, 117, 118, 119, 121, 129, 131, 132, 136
public health, 129
public interest, 126
Puerto Rico, 88, 97

Q

qualitative differences, 92
quality improvement, vii, viii, 2, 40, 41
quality standards, 126, 128, 135
Queensland, 100

R

rainfall, 128
recommendations, 22, 49, 66
recovery, 15, 34, 37, 55, 56, 69, 70, 76, 78, 103, 116
recovery plan, 37
recreation, 9, 128, 136
recreational, 31
recycling, 130
regrowth, 72
regulations, 1, 2, 12, 14, 53, 67, 69, 120, 121, 129, 132
regulatory requirements, ix, 121, 123, 127, 132
rehabilitation, 118, 119
reinforcement, 73
relevance, 113
remediation, 131
reproduction, 104, 105, 106, 109
requirements, 1, 2, 15, 19, 27, 47, 49, 69, 106, 108, 117, 125, 126, 128, 130, 131, 132
resident biota, vii, viii, 3, 78, 104
resolution, 20, 22, 94
resources, 4, 5, 10, 16, 17, 18, 19, 22, 26, 27, 28, 44, 47, 53, 60, 81, 94, 97, 109, 112, 114, 119, 133
respiration, 28
response, viii, ix, 3, 4, 5, 8, 9, 11, 13, 15, 17, 18, 21, 24, 25, 26, 27, 28, 30, 33, 38, 59, 60, 76, 80, 84, 88, 92, 93, 95, 96, 103, 104, 108, 109, 120
restoration, 11, 13, 15, 16, 19, 26, 27, 39, 43, 54, 71, 72, 73, 74, 75, 76, 80, 81, 82, 90, 108, 116, 117, 119, 126
restoration programs, 13
restrictions, 42, 43
rights, 1
risk(s), 59, 60, 79, 114
risk assessment, 114
river systems, 57, 92, 93
root, 121
rules, 30, 31

runoff, ix, 6, 9, 31, 64, 65, 66, 74, 75, 91, 124, 128, 132

S

salinity, 90
salmon, 11, 43, 44, 71, 93, 94, 95, 96, 99
saturation, 42, 43
science, 100, 119
sediment(s), 11, 12, 37, 49, 57, 58, 59, 60, 63, 64, 72, 73, 74, 75, 76, 77, 82, 90, 91, 93, 94, 95, 96, 97, 104, 110, 128
sedimentation, 6, 71, 74
seeding, 73
sensitivity, 5, 8, 10, 33, 79, 88
septic tank, 64
services, 2, 90, 100, 113
sewage, ix, x, 41, 123, 124, 127, 130, 132, 136
shape, 124
shellfish, 9, 104, 127
shelter, 75
shoreline, 87, 90, 91
showing, 70, 96, 127
sludge, 69, 132, 136
social indicator, 116
society, 119
soil erosion, 64
soil type, 49
specialists, 27
species, vii, viii, ix, 3, 4, 5, 8, 10, 21, 26, 27, 28, 29, 31, 33, 34, 42, 43, 44, 46, 50, 51, 55, 56, 58, 61, 65, 68, 70, 71, 72, 73, 74, 77, 79, 82, 84, 85, 88, 90, 97, 100, 104, 106, 107, 108, 118, 121
species richness, 72, 108
specifications, 135
spending, 130
spreadsheets, 20
Spring, 101
stability, 75
stabilization, 73
staff members, 19
stakeholders, viii, 4, 22, 26, 38, 67, 82
statutes, 2
statutory provisions, 1
steel, 129
storms, 91
stormwater, vii, viii, 3, 6, 10, 13, 64, 65, 66, 75, 83, 85, 128, 131
stress, ix, 4, 5, 9, 13, 17, 18, 24, 25, 26, 28, 31, 36, 37, 38, 50, 51, 53, 57, 58, 67, 85, 88, 91, 92, 93, 95, 96, 108, 109, 110, 111
stressors, viii, ix, 3, 4, 5, 6, 8, 11, 13, 15, 24, 25, 26, 27, 28, 29, 30, 33, 34, 37, 38, 42, 53, 54, 57, 58, 59, 60, 65, 86, 87, 90, 91, 92, 93, 94, 95, 102, 108, 121
structure, viii, 3, 5, 20, 26, 30, 31, 34, 42, 43, 45, 52, 87, 88, 104, 107, 108
substrate(s), 37, 54, 56, 79
surface area, 98
surface waters, vii, viii, ix, 3, 7, 9, 45, 74, 79, 105, 120, 123, 124
surfactants, 68
surrogates, 100
survey design, 22, 118
survival, 104, 107
survivors, 28, 109
sustainability, 113
synergistic effect, 5

T

target, 13, 15, 19, 22, 27, 53, 60, 64, 65, 66
taxa, 5, 20, 27, 28, 30, 31, 46, 50, 51, 53, 60, 68, 70, 72, 73, 74, 75, 77, 84, 85, 88, 94, 95, 108, 109, 110
taxonomy, 118
technical assistance, 125
technical support, 17, 18
techniques, 107
technology(s), ix, 15, 105, 123, 126, 127, 128, 130, 131, 132
temperature, 8, 37, 43, 57, 58, 59, 60, 62, 84, 85, 91, 93, 94, 95, 96, 108, 111
test data, 46
testing, ix, 4, 20, 30, 39, 68, 112
threats, 83
time frame, 77
time periods, 86, 89

Title I, 126, 130
Title II, 126, 130
Title V, 126, 131
Total Maximum Daily Load, vii, 2, 12, 13, 60, 63, 64, 65, 74, 75, 83, 99, 111
toxic effect, 8, 68, 109
toxic substances, 120, 127
toxicity, ix, 4, 8, 13, 19, 33, 37, 38, 67, 70, 77, 105, 109, 111, 112
toxin, 33
trade, 2
training, 134
traits, 33, 106
transportation, 49, 81
treatment, ix, x, 5, 9, 15, 36, 41, 43, 63, 67, 69, 112, 123, 124, 126, 127, 129, 130, 131, 135
tumors, 27, 28
turnover, 109

U

U.S. Army Corps of Engineers, 79, 101, 132
U.S. Department of Commerce, 99, 101
U.S. Geological Survey (USGS), 6, 111, 133
uniform, 90, 105, 134
United, 14, 24, 78, 100, 107, 120, 130
United States, 14, 24, 78, 100, 107, 120, 130
updating, 18
urban, ix, 11, 31, 55, 64, 74, 124, 128
urban areas, ix, 124, 128
USA, 37, 97
USGS, 6, 11, 111

V

validation, 20, 80
variables, viii, 3, 85, 88, 105, 120
variations, 36
vegetation, 71, 73, 82, 105
velocity, 75
vision, viii, 3
vulnerability, 85

W

Washington, 2, 98, 100, 101, 102, 103
waste, 68, 127, 135
waste treatment, 135
wastewater, x, 8, 9, 36, 43, 63, 67, 68, 69, 109, 112, 124, 126, 127, 129, 130
water chemistry, 13, 66
water quality standards, viii, 3, 8, 38, 40, 41, 44, 46, 48, 54, 64, 75, 78, 79, 89, 105, 112, 113, 117, 126, 127, 130, 132
waterbody, viii, 3, 5, 7, 9, 12, 14, 18, 19, 20, 21, 22, 24, 26, 27, 28, 31, 33, 40, 41, 43, 45, 46, 47, 48, 49, 55, 57, 64, 65, 74, 76, 79, 87, 88, 89, 90, 91, 92, 104, 105, 106, 107, 109, 116, 128
watershed, 13, 16, 17, 20, 22, 31, 38, 40, 41, 45, 48, 49, 53, 60, 61, 62, 63, 64, 65, 66, 71, 75, 79, 82, 86, 87, 89, 91, 92, 93, 94, 103, 107, 114, 120, 135
waterways, 9, 57, 72, 126, 129
web, 51
welfare, 129
well-being, 111
wetlands, 4, 11, 24, 27, 38, 92, 132
wilderness, 45
wildlife, 9, 44, 64, 74, 127
worms, 28

Y

yield, 118, 119